LEADING THE
LEAN INITIATIVE

LEADING THE LEAN INITIATIVE

Straight Talk on Cultivating Support and Buy-In

JOHN W. DAVIS

Productivity Press • Portland, Oregon

Additional copies of this book are available from the publisher. Discounts are available for multiple copies through the Sales Department (800-394-6868). Address all other inquiries to:

Productivity, Inc.
P.O. Box 13390
Portland, OR 97213-0390
United States of America
Telephone: 503-235-0600
Telefax: 503-235-0909
E-mail: info@productivityinc.com

Cover design by Rochelle Mallett
Page composition by William H. Brunson, Typography Services
Printed by Malloy in the United States of America

Library of Congress Cataloging-in-Publication Data

Davis, John, W., 1938–
   Leading the lean initiative : straight talk on cultivating support and buy-in / by John W. Davis.
      p.   cm.
   Includes bibliographical references and index.
   ISBN 1-56327-247-4
      1. Waste minimization.   2. Manufacturing processes—Waste minimization—Case studies.   I. Title.

TS169 .D39   2001
658.5—dc21

2001019525

05 04 03 02 01   5 4 3 2 1

Contents

Preface

THE INTRODUCTION OF lean manufacturing principles, concepts, and techniques during the last decade has forced most manufacturers to make enormous changes in the way they conduct business. As a result, management and stockholders alike have substantially increased their expectations. In addition, customers have come to expect more, in terms of quality, delivery, and reliability. It has generally become recognized that being competitive no longer means simply staying ahead of the existing competition. Today, the challenge almost every business operation must face, at some point, is how competitive it is on a global scale.

I addressed these issues in my first book, *Fast Track to Waste-Free Manufacturing*, and pointed out that the challenge for the average plant manager is no longer to guide a ship on some well-defined course through what, in the past, were reasonably calm seas. Today the journey requires navigation through frequently uncharted waters and often treacherous conditions. Therefore, most would agree that how prepared the captain of the ship is in leading this effort has a great deal to do with the overall success of the journey.

What's going to be critical for the future, of course, are seasoned plant managers who are willing to change and who are filled with enough knowledge and conviction to keep things moving in the right direction. Helping those taking their first steps in a plant management position will be equally important in order for them to gain respect quickly and, therefore, the active support of the workforce, which, in turn, establishes an environment that is truly conducive to continuous improvement.

With this in mind, I decided to devote an entire work to this matter. Initially I intended to gear it exclusively toward the new plant manager. However, as this work progressed, I realized how much of it applies to almost any managerial position, in almost any business, and how much it can help those taking their first steps in a leadership role.

While being an effective leader is difficult and rests primarily on individual talent and ability, some common themes apply to a plant management position. The purpose of this book is to examine these and to offer some advice and counsel on how to deal with them.

In any profession, a time comes when individuals have to prove they have what it takes to get the job done. As it applies to a management position, some specific indicators of success or failure can be applied: the plant did or did not meet its prescribed budget, did or did not achieve production schedules, and is or is not keeping customers satisfied. However, unlike other positions outside the managerial ranks, silent indicators also exist, which are neither stated nor documented but are nonetheless critical to a manager's success or failure. These particular performance indicators tend to be much more subjective in nature and come as a result of those in higher places making perceived judgments regarding a manager's effectiveness. Here, special considerations come into play. One such example would be that while the manager is getting the job done, he or she could seriously be alienating the workforce.

As a result, the new manager (especially the new plant manager) must quickly be perceived as capable of doing the job and as having the ability to deal effectively with the workforce because, unlike the sports world, rookie plant managers do not spend their first year on the bench learning the ropes. Typically, they are thrown headfirst into a sea of problems and difficulties, and they must come up treading water like a professional. Otherwise, they simply drown in the process.

At some point in almost every manager's career, he or she will face managing unpopular change. This typically sets up the ultimate challenge, which can often make or break a reputation. Thus, how he or she behaves will also say much about the ability to lead the workforce effectively.

Managers should pray this particular challenge is of their own making and not others' because the initiative that stirs up resistance is more manageable if a manager initiates it. As a result, he or she will be more enthusiastic, if not emphatic, about seeing it through to completion. The worst thing would be for an initiative to be perceived as something the manager had been instructed to do because this can damage what every good plant manager is building: a reputation as a strong leader with convictions about what's best. Therefore, the key is to maintain an air of ownership over any change the manager is charged with leading, whether or not it's his or her idea.

However, before implementation, you must evaluate the potential consequences surrounding change since managing change effectively requires confidence as to the likely reaction. This doesn't mean the manager can predict the reaction, but if the manager understands what will occur and how to deal with it, he or she will be in a better position to lead change effectively.

Unfortunately, until new managers gain the reputation of being able to do the job, they are at best limited in what they can expect the workforce fully and actively to support. If nothing else, new plant managers must, consciously or otherwise, convince others of their leadership ability. Unfortunately, few plant managers receive special training or counseling prior to assuming their post. However, if they did, the training would probably be structured as a personality primer. Further, it probably wouldn't cover how to manage the job's activities. Ideally, training would advise and direct managers in the handling of specific issues and situations all of them will face, regardless of the type of business or the particular mode of operation:

- How to go about holding first meetings with staff and employees
- How to develop and effectively earmark a plan for the operation
- How and when to listen carefully
- How to help others get the best out of their careers
- How to apply yourself to a company initiative you don't fully support
- How to work with the union, customers, and other functions
- How and when to buck the system
- How and when to get involved in personnel issues
- How to provide quality time with and for employees
- How to show employees you care for them and the operation
- How to react when surprises and crises arise
- How to implement a new initiative which you desire to be successful

This checklist doesn't cover every topic addressed in this work, but it does list some of the important subject matter. Also included, and of utmost importance as it applies to managing inside the manufacturing area, is some strong emphasis on how to lead the lean manufacturing initiative. A primary focus will be on the obligation to insure the process extends beyond the production floor and into other major functional areas.

This work isn't intended to address the mechanics of implementing lean or waste-free concepts and techniques. However, you cannot lead a successful waste-free manufacturing venture (or any other lean manufacturing approach) if you do not gain the respect, support, and trust of the workforce and management. This is true for the new plant manager who has been thrust into the position during a significant change initiative and even more so if he or she is responsible for leading one from the offset.

Armed with the ability to understand how to lead effective change and with some specific information related to the issues noted above, a new plant manager could in essence use this work as a handbook. For those already in leadership roles, this work could serve as positive reinforcement for what you may already be doing and could help you more successfully manage things that are certain to come your way.

It is my intention to walk you through some of the more important steps associated with getting started in an effective manner. Since my leadership experience was in manufacturing I will, of course, structure this around a plant manager scenario. Again, remember that much of this can apply to almost any middle- and upper-management position.

This work has two distinct parts. The first part, as already mentioned, deals with helping new managers (and specifically new plant managers) to get off to a proper start. Further, it is designed to help them look over the horizon, so to speak, for a glimpse at what will assuredly be coming their way and, most important, what they can do about it.

The second part deals with the role of plant managers (or, in some cases, those in other key positions) and the responsibility they bear in taking the lean initiative to its ultimate level of accomplishment. As mentioned, this primarily deals with insuring it is driven into the order entry and delivery system with the same enthusiasm that is used on the manufacturing shop floor (to name some of the areas that should consider reducing waste.)

All of this, of course, is geared toward one key objective: to make your business the unquestionable leader when it comes to customer service. The truth is, if you lead the competition in this area, you will soon lead in customer satisfaction, market share, and overall profitability. In addressing the above, I will augment the subject matter text with a continuing story, dispersed throughout the book, in order to reinforce the important points.

This story is about the challenges facing a plant manager, Jim Warring, and his frustrations, failures, and accomplishments. What you will find is that being a successful manager isn't a matter of perfection. Instead, it's more about responding effectively to the unknown and the unexpected because, in the manufacturing world, the unexpected is usually the rule and the unknown is usually what serves to make or break reputations.

INTRODUCTION

LEADERSHIP! AT ITS HIGHEST LEVEL, it has been the source of great human suffering and indignity imposed upon millions, but it has also been the catalyst for majestic historical achievements that have served to both advance and enrich mankind throughout the ages.

Most people admire leadership, but few seek it. Achieving it is generally as much the result of appropriate timing as talent, ability, and desire. It is a wonderful gift as well as a sizable burden and is, in essence, the soul and armor of change.

It has been said good leaders are born, not bred. To some extent, this is true. However, as the role of leadership applies to the world of industry, leaders are appointed more often than predestined. Consequently, leadership has to be learned. Most often, this has been done under the aegis of various mentors along the way and more often through the School of Hard Knocks.

There are many available books on the market that deal with implementing the basic tools and techniques associated with the principles and concepts of lean manufacturing. Conversely, most consultants and professionals have not fully recognized that a hard as well as soft side exists to the equation for successfully leading such a process.

The hard side deals with the tools and techniques mentioned. Frankly, once you surmount any initial resistance to change, the hard side is easy. The soft side, on the other hand, is something that has not been adequately addressed and, thus, one of the purposes of this work. This side deals with the job of leading such a process, while handling the basics of the business to keep the process moving forward at a reasonable speed.

To begin, we need to be totally frank about lean manufacturing. Even the name implies the end result that most unions and employees fear: the loss of jobs and, more important, long-term job security. Though any good lean manufacturing effort can weed out an overstaffed workforce, the benefits will be a more competitive operation which leads to increased business, employment, and thus overall job security. When change first occurs, the interim period is generally a trying time for everyone involved. Therefore, leadership must properly focus on the objective while dealing with the more classical aspects of running a business. This, of course, includes the often growing frustrations of employees during such a substantial change in business conduct.

Based on my personal experience in plant management and lean manufacturing (with a process I called waste-free manufacturing), I recognized that the soft side of the equation is the more difficult of the two to manage. No matter how important enthusiasts of lean manufacturing feel about the need for becoming lean, at all costs, the normal course of business prods on and can often become a serious stumbling block. Therefore, you must establish a unique balance between the need for change and handling the business's day-to-day aspects making the difference between long-term success or failure. Establishing this balance, of course, is an operation's plant manager's key responsibility.

As with life, manufacturing operations have a regenerative cycle. New ideas and concepts are born and nurtured as others pass away. But out of all this hubris comes constants. Unfortunately, these do not conveniently go away when you decide to change the operation's production mode, as is the case with any serious lean manufacturing initiative. Therefore, they have to be dealt with in an effective manner.

Though this book is directed at plant managers (more specifically those relatively new to the job), it can be of unique benefit to almost anyone taking on a leadership role, regardless of position. What it serves to address touches the heart of effective leadership: competently dealing with the unexpected. This is something every leader will face in his or her career and on more of a frequent and recurring basis than most believe.

Obviously, if everything occurred as expected, we would not need a leader. Without the unexpected and unknown, a business could practically run itself. The reality is things do not always go as expected or anticipated. That would be the exception rather than the rule. Unfortunately, some plant managers act more like robots than living, thinking human beings. When the unexpected does occur they don't know how to react because they operate on the basis of what they have been programmed to deal with. Thus, they subsequently error in their duty or, in the worst case, professionally self-destruct.

What manufacturing and the business world need are managers who have learned to expect the unexpected, who go to work each day knowing and anticipating it is going to happen. They stay organized under pressure because they predict it and they view it not as a run of bad luck but rather as a leader's responsibility. They are so prepared for it that they are disappointed when it doesn't happen, for it is a challenge they welcome and respect. The typical manager, instead, cringes at the thought of the unexpected and panics and/or overreacts, when it does occur.

This work focuses on two important aspects of an effective manager. The first, as noted, is anticipating and dealing with the unexpected. The second is how to go about building and maintaining the proper *perception* of others. How a manager's qualifications and accomplishments are perceived is the catalyst for what his or her reputation becomes, especially true for the plant manager. Unfortunately, while we would like to believe results measure a manager's ability, this typically isn't the case. Even outstanding results will be frowned upon if the perception is it was achieved through something less than a genuine concern for the company and the workforce. If the results are viewed as self-centered or self-promoting, the manager will lose the long-term favor of many within the organization.

But allow me to stress that managers will not be perceived as either capable or successful if they cannot attain reasonable accomplishments. That aside, however, how others look upon a manager's method and manner for conducting business is crucial to getting the job done in a manner that supports both teamwork and the workforce's continuing enthusiasm.

What I have discovered is some predictable points in a plant manager's tenure during which this perception occurs, regardless of the business or product. So, follow me, as we examine these. On the way, you will learn from Jim Warring's experiences, a young plant manager taking his first steps in the role. As you will see, the unexpected and unknown proceed to be something he never fully anticipated. From this will come a wealth of insight and understanding for those aspiring to, just entering, and experienced in such a position. In the process, three important truths will become evident:

1. Leading the waste-free initiative—or any other lean engineering effort for that matter—will be made difficult due to the day-to-day distractions a plant manager must deal with.

2. The key to maintaining focus and achieving effective results rests on how you approach certain facets of the job in setting the stage and establishing a vision for the future.
3. The waste-free initiative has to be energetically driven far past the manufacturing arena, into the total business chain, if it is to take its rightful place as a waste eliminator and, thus, as a true competitive advantage.

1

Setting the Stage

UNDER THE GUISE OF MANY NAMES, the world manufacturing arena has been in the process of a significant transformation over the past two decades; with a pronounced increase in activity during the last decade. This transformation has theoretically been tied to changing the principles, concepts, and techniques fundamental to manufacturing's job performance, with less waste and more efficiency. It all started with the Toyota Production System (TPS) well over four decades ago; more recently, in the United States, it has come to be popularly termed lean manufacturing—or, in some quarters, lean engineering. In the most basic sense, it reverts manufacturing from an untidy and disorderly operating unit (which in most cases is waste-filled and inefficient) into a highly disciplined and more proficient business function.

Unfortunately, however, America still lags behind a number of competitors in this race, specifically Japan, which has taken its world class manufacturing to unparalleled heights. The question becomes why and the answer lies in two fundamental shortcomings that permeate U.S. business. The first is big industry's inability to remain focused and to resist every new concept offering a unique advantage over the standard or accepted way. This inability has created an overload of programs that have not attacked the root problem and helped to move U.S. manufacturing to a true leadership position. The other shortcoming is a loss of some appropriate power and control at the manufacturing level. Consequently, this has created a growing decline in leadership effectiveness in manufacturing because power and control are no longer considered the path to expedient career growth and advancement.

The second shortcoming has come because more operational control was shifted from manufacturing to the financial, quality assurance, safety, and environmental sectors (among others), which were viewed as support functions to manufacturing. Therefore, U.S. manufacturing no longer carries the power and influence it once had over its destiny.

As with most changes in an established power base, this shift has had good and bad effects. As a benefit, U.S. manufacturing operations currently produce better quality, on average, than they did in the past, and they are unquestionably much safer and healthier places to work. As a detriment, much of the realized profit improvement has been achieved through significant downsizing and/or moving manufacturing operations out of the United States to countries with lower labor and overhead costs.

So, how would I rate the general progress of U.S. industry in its efforts toward excellence in manufacturing over the past two decades? I would say only fair at best.

The automotive sector in the United States indicates how American industry and manufacturing, in particular, score on the scale of world-wide competitiveness. This industry has instituted countless concepts theoretically tied to a world-class approach to manufacturing; insisting its suppliers follow extremely strict guidelines to improve their systems, procedures, and production processes. However, this misplaced emphasis has seriously restricted how and when its suppliers can make changes, dampening America's potential from fully achieving and maintaining a true, globally competitive status.

For reinforcement, review a *Wall Street Journal* article by Robert L. Simison and Joseph B. White (May 2000). The *Arkansas Democrat Gazette* partly reprinted the article (May 7, 2000), and the headline read:

"Detroit driven to raise quality as U.S. autos still trail Japan's best"

Here are some noteworthy comments from this article:

"With U.S. sales booming, top executives at General Motors, Ford Motor Co., and Daimler-Chrysler AG's U.S. operations played down quality in recent years and insisted the gap with the Japanese and Germans was narrowing. Quality was no longer an issue, they said: It was a given. But despite a massive overhaul of manufacturing and engineering systems over the past two decades, a spate of rankings and reports show U.S. cars still lag behind the best from Japan and Europe. And complacency about quality has started to give way to something closer to alarm."

"Strategic Vision, a consulting group based in San Diego, last month named the top vehicles in 16 market segments based on its Total Quality Index. Owners rated craftsmanship, reliability, driving performance, and their emotional response to the vehicle. Volkswagen, despite small market share, won five categories, including a tie. Daimler AG's U.S. products won three categories; all GM vehicles, only two; and Ford, only one (and that was part of a three-way tie.)"

I strongly contend that the search for enhanced product quality and the full achievement of lean manufacturing are identical. However, many will tell you this is not the case. In their minds, lean manufacturing is a process focused on a production system aimed at improving manufacturing efficiency and, therefore, costs (through productivity and various throughput gains.) But while this has had some quality enhancing aspects, it simply did not go far enough. As a result, they insist, other partnering programs and processes are required.

As I previously mentioned, the facts will show U.S. based manufacturing has a long standing history of mirroring the automotive industry's lead in just about all matters associated with the way manufacturing is conducted. This has not created a panacea for world leadership in the U.S. manufacturing arena. If this were so, we would not be seeing the fast exit of manufacturing to our southern borders and elsewhere in the world.

We still desperately lag in this race and this can be strongly attributed to what appears to be an undying need to always strive to create a better wheel. For example, working hard to follow the simple yet proven concepts of the TPS was considered to be too simple to be effective. In turn, the automotive industry established its own reinforcement processes. A perfect example is QS9000, which has reduced the importance of continuous, unrestricted change; it has instead placed a strong emphasis on a philosophy of good business practices based on detailed documentation and follow-up procedural practices. QS9000 is basically counter to the fundamental principles of the TPS because it demands extensive documentation in order to proceed with change and thus delays implementation of ideas and concepts and positive benefits.

As a result of layers of documentation, ensuring an established procedure (followed to the nth degree) has become more important than making continuous and effective change. In fact, the new systems' associated policies has made change harder and more difficult.

Many had raised the self-perpetuating argument that unrestricted change to manufacturing processes has led to serious quality problems and, in fact, some unfortunate recalls. Though this may be true, the same can be said for design engineering errors and we do not see damaging restrictions being placed on engineering. The answer in the latter case has been to learn from experience and then proceed. In the case of manufacturing, however, the answer has been to discourage not only any change (once a process has been deemed acceptable) but to make change almost impossible. Even worse, when changes result in minor problems, the practice has been to point fingers and/or chastise those in charge. As a result, the price of U.S.-designed/manufactured automobiles has continued to rise dramatically, while quality—in the consumer's eyes—is viewed as sub-standard compared to Japanese auto makers and others.

This lack of change has caused a lack of manufacturing parity with some of our major competitors (specifically, Japan) when it comes to ingenious changes that continuously enhance the overall manufacturing process. However, how does all this relate to strong individual conviction and leadership at the manufacturing level, which I strongly tout as being absolutely critical to the future of American manufacturing?

First, understand that TPS makes one simple, yet powerful assumption: Ongoing enhancements, or Continuous Improvement, to its manufacturing processes are a competitive must. For this to happen, few restrictions are placed on when and to what degree change can be made. Everyone at Toyota realizes that nothing worthwhile is totally risk free. Therefore, good intentioned mistakes are applauded rather than rebuffed and/or criticized.

Toyota's system of production recognizes change must be continually encouraged, supported, and reinforced. Unfortunately, most American industries have not learned to delegate, empower, and encourage their employees to seek out and make continuous change. Initial design plays the principal role in quantifying a product's quality and reliability. If most products are designed to meet a specific quality standard (a quality level the consumer is willing to pay for), then the major driver for applied quality and reliability rests within the manufacturing arena. Restricting continuous improvement to the overall manufacturing process in any way then becomes an obstacle to success. Unfortunately, this has happened in American manufacturing and is the major philosophical difference between the more successful Japanese and our lagging and rapidly disappearing U.S. manufacturing base.

Again, plant managers must understand how to stay focused on this matter regardless of the certain countless distractions and disruptions: That is what this book will address. Every plant manager, no matter what the particular industry or market, will face the unexpected and the unknown, which can often be the most serious stumbling block of all. All younger plant managers and others in manufacturing hold a key responsibility for bringing the kind of ongoing and extensive change that American industry so desperately needs.

U.S. industry, in following the automotive sector's general lead, must recognize and create an acceptable balance for continuous improvement at the manufacturing level. If not, we will continue to be another also-ran in the global race for manufacturing dominance.

For those non-plant managers who comprise most of our readers, the obvious question is how can this book help me? The answer is that every worldwide business position (especially in manufacturing) will face the same basic challenges a plant manager faces. The unforeseen and the unexpected are the enemies of every profession. Only when most employees recognize and maintain focus on the proper principles, concepts, and techniques, will U.S. manufacturing take its rightful place as a world leader.

Finally, U.S.-based companies must appreciate that lean manufacturing principles have to extend beyond the manufacturing arena. At some point, the other key business functions must adopt them, or the advantages gained at a manufacturing level cannot be leveraged as a competitive weapon.

Cutting manufacturing process and lead times serves little purpose if such improvements are not incorporated into order and delivery systems and, thereby, fully utilized for a competitive advantage. Without this incorporation, the lean manufacturing achievements are window dressing and do little for the business. The improvements must somehow rise to where customers see them as a unique advantage for their particular needs, for when the customer sees such an advantage, then a competitive edge has been established.

The more serious stumbling blocks facing U.S. industry are the distractions that can and do block a solid commitment to continuous improvement. Therefore, the manufacturing level job never gets done and the benefits of a sound waste-free atmosphere do not materialize to the company's advantage.

As I have mentioned, we will address the unexpected and the unknown, which can be our chief enemies. They can make us cowards and, certainly, can drain away needed energy and appropriate focus on the sizable task facing America's manufacturing leaders. When we fully understand, we will recognize (if not forecast) their coming and be prepared to combat them. The Warring Saga clarifies this point where much emphasis is placed on outlining the common occurrences distracting this young plant manager from his goals and aspirations. Had he learned to anticipate and, therefore, expect the unexpected, he would have dealt with it more effectively.

CLEARLY UNDERSTANDING THE KEYS TO SUCCESS

Before we examine the new leader's first steps for gaining support and cooperation, we must define the keys to success as applied to leading the lean manufacturing initiative. Success isn't how well one knows the tools and techniques and/or how active they may be in assisting with the process implementation. Success principally boils down to having a clear understanding of the initiative's depth and scope in order to be effective. This naturally has to be coupled with the desire and will to stay focused on making it happen.

Without first having a clear understanding of depth and scope, you will venture into a significant change process without understanding how far you need to go or when you have arrived. *Fast Track to Waste-Free Manufacturing* addressed the journey's preparation and covered the importance of understanding and identifying your direction, arrival time, and passengers going along for the trip.

But in more specific terms, this understanding centers on three key factors. The first is the drivers behind such dramatic change at the manufacturing level. In other words, you must clearly understand what has caused this pronounced attention on re-engineering manufacturing. Certainly, the question does arise as why begin with manufacturing? Why not sales, marketing, engineering, etc., within the business organization?

The second factor pertains to the process selected to correct the company's decline. Generically, the common approach has been referred to as lean engineering, with different names and styles. However, most processes have a common theme that requires them to follow the principles blindly set forth in the TPS. waste-free manufacturing is not in that category. Consequently, fast and effective change has been less than stellar for most U.S.-based manufacturing operations. While some change for the better has been made, a problem exists with maintaining the proper disciplines and appropriately focusing on continuous improvement.

The third factor—where most operations fail miserably—evolves around the appropriate linkage of the improvement process to the total business chain. The gains made at a manufacturing level do not extend into the chain of functional events that (between order and delivery) serve to form the customer's most crucial impression of the company.

The Drivers

To understand this matter, we should start by examining these drivers. One of the better works on this particular subject is *Kaisha—The Japanese Corporation*, by James C. Abefglen and George Stalk, Jr. (Copyright 1985, Basic Books, Inc.) The authors begin by pointing out:

"This book is about the business corporations of Japan—the Kaisha—which now play key roles in the world business arena. We have sought to describe and explain the competitive behavior of Japan's companies to those Western businessmen whose own competitive decisions and actions require a more complete understanding of the Kaisha. This is not another explanation of Japan's economic success. It does not deal with miracles, or conspiracies. It is the author's hope that a more analytic and less emotional discussion of competing with Japan's companies will be of interest and value to the Western business community."

(Preface, page IX, paragraphs 1 and 2) "The dominant competition for a great many companies of the world today is from the Japanese, who are increasingly setting the pace of competition throughout the markets of the free world. The rapid development of Japan's companies, or Kaisha, is startling. They have grown from the debris of a lost war to take world leadership positions in a surprising number of industries." (Chapter 1, page 3, paragraph 1)

It seems miraculous that a country weak in natural resources has so influenced the worldwide conduct of business, indeed a sizable achievement. Although, some 15 years after Abegglen and Stalk's book, Japan's economy has since slipped somewhat, the country still is and will continue to be a tremendous force in the world economy.

Possibly more important (and less touted in the United States than the TPS) is Japan's overall business philosophy. What I'm referring to is their business drivers, outside of some rather startling manufacturing improvements. In a nutshell, these drivers unerringly focus on substantial business growth.

As Abegglen and Stalk point out, growth revolves around the "competitive fundamentals chosen by the successful Kaisha." (Chapter 1, page 5) These fundamentals are the following:

- A growth bias
- A preoccupation with the actions of competitors
- The creation and ruthless exploitation of a competitive advantage
- The choice of corporate, financial, and personnel policies economically consistent with all of the above

In setting the stage for what I will clarify, I will quote two excerpts from their book pertaining to growth bias:

"The strong bias toward growth of successful Kaisha is closely linked to their desire to survive. These Kaisha have been built in a rapid growth economy. They have been witnesses to the fate of companies that failed to grow faster than their competitors. For example, in the late 1950s Honda increased its production 50 percent faster than demand called for to displace Tohatsu in less than five years as Japan's leading motorcycle manufacturer. Tohatsu went bankrupt and forty-five other Japanese producers eventually withdrew from the manufacture of motorcycles." (Chapter 1, paragraph 1, page 6)

"Management with a bias toward growth has distinct mindsets which include the expectation of continued growth, decisions and plans formulated to produce growth, and the unfaltering pursuit of growth unless the life of the organization is threatened. Companies with a bias toward growth add physical and human capacity ahead of demand. Prices are set not at the level that the markets will bear, but as low as necessary to expand the market to fit the available capacity. Costs are programmed to come down to support the pricing policies and investments are made in anticipation of increased demand." (Chapter 1, paragraph 3, page 6)

Consider the following questions:

- How many U.S.-based, manufacturing-driven companies have an unerring philosophy for growth aimed at seriously outpacing their competition for market share?
- How many have made vital plans formulated to produce such growth?
- How many have or are willing to add physical and human capacity exceeding the current demand?
- How many would be willing to set prices as low as possible to fill that added capacity?
- How many would program their cost structure and invest accordingly in anticipation of the demand?

The answer is few because, contrary to popular belief, American industry tends to plan its businesses more conservatively than the Japanese do. Even when a team of American design, sales, marketing, and manufacturing people want to aggressively grab market share—when planning a new or revised product or service—the standards associated with justification won't allow this to happen. Therefore, no one plans on significant growth, no one expects significant growth, and most often it never happens.

Now, to answer why so much emphasis has been placed in American industry on improving manufacturing, we only need look at Abegglen and Stalk's observations. Manufacturing must take the lead role in continuous improvement activities to enhance product quality and delivery and to lower operating costs. However, this hasn't happened in the United States for the same reasons it became a competitive weapon in Japan. In Japan, the need was driven by prices being slashed to fill added capacity and that substantial capital investments had anticipated this.

The best way to describe this philosophical difference is that in the West, the principal focus has been on cost reduction and, therefore, expedient profit improvement. When this cannot be achieved aggressively enough in a domestic operation, through some sort of lean engineering, most companies look at relocating manufacturing (Mexico, for example), where a lower labor and overhead base exists.

In the East, the principal focus has been on expedient processing and, therefore, the ability to deliver when substantial gains in market share occur primarily as a result of setting the lowest market prices. This long-term approach has served the Japanese well establishing greater overall profits and market share.

While many U.S. companies have made great efforts to implement the tools, concepts, and techniques of TPS inside their manufacturing operations, they have not had the same urgency to change outside manufacturing. In addition, they've been less willing to change their business strategies in order to take advantage of competitive manufacturing gains. Even more important, they have had little desire to apply the lean manufacturing tools, which have improved manufacturing to other key functions, such as Sales, Marketing, and Product Engineering.

An analogy would be someone energetically building a high-powered race car then limiting it to a small, oval dirt track. The true power and performance of the vehicle would never be fully realized or appreciated. This is what U.S. industries are doing—in many cases, spending fortunes to bring aboard consultants and applying their own internal resources to an endless array of manufacturing training. However, because most industries don't feel obligated to put their process on a high-performance "track," it never gets the chance to run beyond a hard idle.

This is such a shameful waste and one that is costing U.S. industry jobs and longevity in the markets. Since the so-called new economy has taken up the slack and re-applied so many of the manufacturing jobs lost over the past two decades, no one seems to notice or care much. Still, when we lose enough of our hard industrial base to become almost totally dependent on a service-oriented economy, we will be in trouble as a nation.

The Process and Linkage Factors

As Abegglen and Stalk outlined, if a lack of growth bias loss has reduced U.S. industry's competitive parity, then the reduction in the scope of lean engineering is a key factor, as applied to linkage to other areas of a business.

Most Western operations have not taken lean engineering as far as the Japanese. One of the primary reasons for this has been our lack of emphasis on extending the continuous improvement process into other key business areas outside of manufacturing.

Oliver Wight Limited Publications, Inc., developed *"The Proven Path"* (copyright 1989) handbook, which deals with what the authors describe as the integrated approach to MRP II, JIT/TQC, and DRP. They pointed out the following:

"Usually the seeds to JIT/TQC first develop in manufacturing, but quickly we see that constraints to responsiveness exist throughout the company, not just in manufacturing. Because JIT/TQC will effect the entire company, it is essential to obtain support from all functional areas.... To progress past the pilot stage and achieve a breakthrough for the entire company, the effort will eventually require active participation by all areas." (Page 34, paragraph 2)

Though written in 1989, what was pointed out still holds true today. In essence, the standard improvement process in U.S. industry usually goes something like this:

1. The company has an urgent need to improve profits and decides to change its manufacturing operations to take advantage of lean manufacturing concepts and techniques. (After all, everyone else is doing it and it is the latest fad.)
2. The company readies for this through special communications and then proceeds to train the workforce with external consultants and/or the placement of needed internal resources.
3. The company improves its manufacturing operations. Downtime, scrap, rework, and obsolescence drop. Manufacturing lead times and the quality of workmanship improve.
4. The company decides not to or does not strategically drive the same improvement process beyond manufacturing into the other key business functions.

5. Customers see little, if any, difference because prior to the improve-ment process the quality of workmanship was basically "inspected" into the delivered product. Since the process was not extended past manufacturing, standard lead times for the customer were not revised or communicated. Though the company has a nicer looking and somewhat more cost-efficient manufacturing operation, it real-izes little improvement in its market share and, therefore, its long-term profit potential.

6. The company, to enhance the profit picture further, considers mov-ing manufacturing operations to an area with a lower labor and overhead cost base (Mexico, for example).

Does any of this sound familiar? This practice is more common than you think.

Regarding the chosen process, in America, we have not been any-where near as creative and innovative as we used to be. Instead, we have decided to take the conservative path, which calls for us blindly to follow the rules (principles) set forth in TPS. We should be focus-ing on the positive aspects as well as the obvious shortfalls of TPS and revising it to fit the situation. This situation calls for the rapid insertion of the more important tools, while avoiding those adding window dressing. Few U.S. companies are considering this. As a result, this has established the need for other supportive programs and processes, which I will address later. The fact is, we simply are not moving far enough, fast enough, in either the manufacturing arena or elsewhere.

Waste-free manufacturing offers a means of achieving the needed end results because it uses a new set of principles (rules) designed to use the proven tools and techniques where rapid deployment is achieved throughout the manufacturing arena. But it has to go much further than this.

As important is insuring the manufacturing level improvements are linked to the total business chain. Otherwise, the customer sees the same old way of doing business. For example, through the use of various lean manufacturing tools, manufacturing can lower its total cycle time (from order to delivery) as much as 50 percent or better. However, if nothing is done to fit this into the standard order and delivery cycle, then the customer recognizes no improvement.

So, having outlined this phenomenon, what are the true keys to success? They are the following:

1. Take the initiative beyond the manufacturing arena. The process of lean manufacturing must go further than just the shop floor. If this doesn't happen, such an initiative will have little value to the customer, and when it offers little value to the customer, it has even less value to the company.
2. Carefully select the lean manufacturing process driving this initiative as numerous approaches exist. I strongly recommend waste-free manufacturing, primarily because it is a proven concept that provides rapid insertion of practices and fits well with the American way of making effective changes at a manufacturing level and beyond.

Table 1-1. "Sacred" Rules for Leading the Lean Manufacturing Process

1. Be persistent to a fault
2. Avoid distractions
3. Select co-champions carefully
4. Never compromise key principles
5. Remain loyal to the process

However, through it all, you must remember the importance of the manufacturing leader's role and the actions that he or she must assume in making the lean engineering venture a total success. In most cases, this person will be the assigned plant manager. Regardless of the title, this leader should know five sacred rules in carrying out this important mission. Refer back to these as you read through the chapters in this book:

1. *Be persistent to a fault:* To be effective with any lean manufacturing initiative, you must understand you hold two important responsibilities. One is to implement the process throughout the entire manufacturing arena. You must accept no excuses for delays in the application or for any decision to skip an area or department for any reason. Two is to be an active ambassador, who does all within your power, authority, and influence to drive the continuous improvement process into other key business functions outside manufacturing.

2. *Insure the normal distractions of the job do not deter you from the mission:* All business leaders must understand that the unexpected and unknown will be their enemy with any initiative, especially one with the depth and scope of lean manufacturing. You are talking about significant change, the kind that can create considerable fear, frustration, and anxiety for some within the workforce. This can lead to some considerable resentment, anger, and sometimes a downright lack of needed support and cooperation.

3. *Select your co-champions carefully:* If you are going to lead the initiative, then lead it. This means you must fully manage the process but avoid getting too wrapped up in the training and implementation procedures. You would not be able to do this anyway while handling the other standard duties and responsibilities of your job as plant manager. This means you must select and assign co-champions for the improvement process. They will report directly to you and will be your daily voice to the employees. They should fully share your vision about the mission and hold an unerring commitment to seeing it through to what you, as the leader, have established as the ultimate level of achievement. Therefore, carefully select these individuals and quickly remove those who prove to be unfit or unworthy of the challenge they must undertake.

4. *Never compromise the key principles:* The key principles of the improvement process should be considered as the staunch commandments. In other words, while you and your co-champions can be flexible and selective regarding many of the standard tools and techniques (e.g., Kanban and Takt Time), you must never bend on the key principles. In waste-free manufacturing, the four established principles are: Workplace Organization, Insignificant Change-Over, Error-Free Processing, and Uninterrupted Flow. Under no circumstances do you allow a decision to be made and carried forth that would circumvent any of the key principles that apply to the lean engineering process.

5. *Remain loyal to the process:* Belief in the lean initiative is crucial. One's integrity should never be compromised by hollow support from upper management. Be prepared to leave the operation if and when top management support evaporates. Fifty percent (or more) of companies who decide on a lean initiative are neither totally serious about it or willing to be persistent to a fault. Therefore, don't waste your time managing and pushing a process forward

which will not be seriously adopted as a continuing way of life. Too many other good companies could use the talent, expertise, drive, and focus you can bring to bear on the matter.

A final word about the fifth rule, which is especially important if you have firmly decided to make a career in manufacturing. To spend valuable time and energy with a company—neither serious about lean engineering or not prepared to go all the way—can endanger your career. Leaders who are serious about staying in manufacturing should strive to insure that they settle with an organization who sees the need to be globally competitive, through the use of world-class manufacturing concepts and techniques. Otherwise, your talents and abilities, over time, will become obsolete. On the other hand, the first time you run into the slightest resistance, do not pull up stakes and head elsewhere. Some resistance to substantial and expedient change must be expected from management as well as the workers on the shop floor. What I am suggesting, however, is to keep a watchful eye on this issue and make certain that you do not spend precious time and energy with a company with no intention of letting you manage effective change.

2

Getting Started

IT'S YOUR FIRST DAY as the newly assigned plant manager and time to be introduced to your staff of direct reports and later to the workforce. Surprisingly, after all the pre-job advice you've received, no one has come to tell you what you should say or do. You feel totally alone and unsure of your next move. You hesitate to ask for advice for fear of being perceived as inadequate and you rack your brain for the right words to say. You seek something inspiring, without coming across as arrogant or threatening. You want them to see you as being tough, but understanding. Firm, but with a flair for handling difficult and unique situations.

You're new, having been hired from the outside, so you see two tactics you can take at this point. One would be to extend a friendly hello and a comment about the privilege being selected for the job and becoming a part of the team. The other would be let them know you've done your homework, you've researched the company and then elaborate on your plans and expectations.

As you sit next to the vice-president of manufacturing, you see your future direct reports enter the conference room and mill around. You realize most will be sizing you up, examining every word you say and how you present yourself. So, what do you say?

In almost every first meeting with your direct staff, you should choose the first option. Why? First, and foremost, remember you are not in charge of this meeting. The vice president, to whom you report, is. Plus, whether or not you like it, you will be giving that first and important impression to your staff and your boss. Therefore, the less said the better.

The second and perhaps more important reason is to be certain you avoid wrong impressions. In such a normally short meeting, do not make a point, for it can be misunderstood and subsequently misrepresented, regardless of people's good intentions.

However, what can happen is your boss will make an initial introduction and then leave you time alone with your staff. He or she may or may not let you know about this beforehand. If it happens, what should you do? The answer is what you would have done if he or she were there. Say hello, show sincere appreciation for being chosen, give some assurance about looking forward to working with them and, most important, let everyone know your next step. Try something like this:

"First, I'm honored to have been given the opportunity to work with you. I'm grateful to have been chosen for the post I've been asked to undertake and I look forward to us working together and setting a course for the operation that's going to assure continued success. Most important, as my staff, I want to give you an idea of what I will say to the workforce in my communication sessions. It's not going to be an oratory on what I have planned or even expect out of our operation. Before I can do that I need to learn more about the issues we may be facing. When that's done, I don't intend to draw up some plans and then let you and the rest of the workforce know about them. I intend for us to manage this operation together, as a team. And, as a team, I would like to take this opportunity to get to know each of you better, so let's go around the table and you can tell me a little about yourself and your background."

This, with an appropriate closing, should satisfy the first staff meeting. These short remarks did and did not say certain things about you, as a manager.

What it clearly said:

- You are going to keep them informed, if possible, because you let them know your next step, e.g., what you plan to say to the workforce.
- You are interested in them by asking they tell you a little about themselves.

What it implied:

- You are not entering the operation with a lot of preconceived notions about what should or shouldn't be done.
- Teamwork is important to you, with the use of us/our rather than me/mine.

What it didn't say:

- You don't have strong opinions and certain expectations of them and others.
- You can't be tough and demanding when the need arises.

I once attended a meeting where a new plant manager was introduced to the salaried workforce and after many years, I haven't forgotten the less than positive impression he left with just one remark. Though he turned out to be successful in the role, he could have been more so had he been more thoughtful about his opening comments. The truth was he wasn't like what he tried to portray himself to be (which as you will see can be a problem in itself). Nonetheless, his comments sent an unnecessary shockwave through the ranks, which later led to some good people quitting.

His comment was: "There are a number of you here today who are not going to make it over the long haul." That comment served to alarm almost everyone; and what I find interesting is that almost 30 years later, it was the only thing that really stuck with me. I saw it then and still view it today, many years later, as something totally uncalled for.

Time has a way of revealing the truth, and perhaps some employees were indeed better off because of his comment. Still, the ultimate nemesis of productivity in any organization, the *one* thing that makes people less effective in their work, is any real or perceived fear regarding job security. All plant managers, new or seasoned, should remember this. Unfortunately, during every plant manager's tenure, the time comes when layoffs and restructuring efforts cannot be avoided, thus job security issues will arise. However, even with the best intentions, to broach this subject when not necessary is asking for trouble.

SOME SPECIAL COMMENTS ABOUT THE EXTENDED STAFF

Almost every business has a *direct* and *extended staff* of managers (staff means the general management team). The direct staff normally reports to the plant manager, and usually comprises department or unit managers in charge of directing and leading specific organizational functions. The extended staff often reports directly to the department managers and usually excludes members of an assigned labor bargaining unit.

Regardless of the specific organizational structure, a team of people usually directly supports the plant or general manager. This team's job is to guide and lead those who make the products and/or provide the services being delivered.

When these two managerial levels coexist (i.e., a direct and extended staff), the new plant manager will usually meet with the direct staff prior to being introduced to the extended staff. Your first meeting with the extended staff should not occur during the same meeting with the workforce but should be done separately. You want to get your extended staff on your side first. If you accomplish this, your ability to make a smooth and effective transition of leadership will increase.

In every operation—particularly with a salaried employee group and an hourly, often unionized, group—a distinct line of authority is drawn within the organization. On one side of this line is the management team, and on the other are those who take daily direction from that team. If you belong to an operation where this is not the case, then you are the exception rather than the rule, because most organizations are structured as noted. However, within the management team are some well-defined authority and responsibility lines, which are often guided by how a plant manager both perceives and decides how to utilize this important resource.

One of the biggest mistakes plant managers can make is not giving appropriate recognition to the extended staff as full-fledged managers. While many may not have direct reports and may serve as individual contributors, they are definitely managers and should be treated accordingly.

Once aboard, the extended staff is expected to manage daily activities with little, if any, specific direction. Many manage crucial projects and most manage their own personal contributions to the operation's goals and objectives. We expect it of them and they normally deliver. However, if they are our future leaders, how can we expect them to be effective in that role if we do not first recognize them and then give them the opportunity to grow as managers? The answer is that we can't.

The highest degree of organizational turnover is within these ranks, and this can be viewed as inevitable as these particular individuals gain experience, grow professionally, and move to new careers. But if you pay close attention, you will notice that most move to positions outside the company. Though people do not want to burn bridges, if you analyzed their exit interviews you would usually find most didn't leave for more money. They left because they were not being fully utilized or substantially challenged.

What a shame for a company to put time and energy into recruiting and training such a valuable resource, only to be faced with constantly replenishing it with new and often inexperienced talent. Equally important is the never-ending cost and impact this has on a company's overall productivity. The issue becomes how we can do a better job of salvaging this important resource. You must understand that few of this valuable team would seek outside opportunities if they believed they were essential management members. Believe me, this is not an oversimplification to a monumental problem.

My experience is that individuals leave from their job to positions of equal or slightly better status. How they view their future advancement opportunities, reinforced by what they say when leaving, is a significant indicator. If you hear: "It's a better opportunity for me" or "I hate to leave, but couldn't pass up the opportunity," the underlying truth could be they feel unchallenged or incapable of advancement or both.

Though they cannot be totally candid and say they didn't see an opportunity for growth or advancement, they will usually send a message. Therefore, listen more to their words than to their method.

The plant manager's role in effective succession planning is one of the more important tasks of senior managers and plant managers. However, most senior managers do not take this role seriously and do not see it as a prime job responsibility. Rather, they view this as something the HR manager and others should lead and direct, as a integral part of their ongoing duties. For the new plant manager, my strongest words of advice are these: DO NOT ALLOW THIS TO HAPPEN.

If you have anything special to tell the extended staff, discuss this subject matter. Let the staff know they are a valuable resource and are essential to the management team. Tell them you will give them an opportunity for promotion, when possible, and you plan to work hard to accomplish this.

When you do meet with them for your introduction and remarks, do it in a special meeting just for this group. Be sure to bring along your staff of direct reports. You do this because most of the extended staff reports to members of your direct staff.

YOUR INTRODUCTION TO THE WORKFORCE

The next step for new plant managers comes when they first speak to the workforce. Organizations go about this many ways, from small gatherings by shift to one mass meeting. I prefer to hold meetings on a shift basis and, if possible, to have the entire management staff at every meeting. The reasons for this are twofold:

- Plant managers usually find distinct differences in thinking between members of the first, second, and, when required, third shifts. The first shift normally contains employees with the highest seniority. Therefore, layoffs and job security can be a greater concern in the latter shifts. As a result, plant managers must typically tailor their remarks to each shift, while maintaining a general theme/focus.
- Though one meeting with the entire workforce works well for an operation with fewer than 50 employees, accommodating a large number of people while being as personable as possible is difficult. Avoid this approach if you can help it. You must have your management team in attendance at all such meetings. It serves to show strength and unity of purpose.

Your first remarks to the workforce will vary. However, unlike your first meeting with the management team, a simple hello is inadequate. For example, a new plant manager has been hired because delivery problems have had a serious impact on market share and profitability. In such a case, especially if the workforce knows something about the situation, they will expect to hear reassuring words and will want to leave feeling the company has made the right choice selecting you. So, the key to getting off to a good start is to conduct such meetings when you are ready to address the operation's more critical issues. Any delay in holding your first meeting with the workforce cannot be weeks in the making, but it can be a few days in coming if required.

Remember some things about your first remarks to the workforce:

1. No one is expecting you to have definitive solutions to every problem. What they hope, however, is for you to recognize the inherent problems and that you can reassure them these problems will be overcome.
2. As the new leader, employees will want to know if you are interested in them and the operation or if your motives appear self-centered. Therefore, sincerity and humility are important, without the appearance of a lack of confidence in yourself.
3. If you have experienced the same problems the plant is now facing, share those personal experiences. Even if you weren't a plant manager, this informs the workforce you have been there before and know about living and working through such situations.
4. Show you're interested in the operation's success and are willing to push yourself to make it happen. Of course, always ask for their continuing help and support.

Regardless of what you tell the workforce, always follow up by doing what you said you would do. Many plant managers forget making specific remarks, which is easy to do in the midst of all the excitement of a new beginning; but the workforce does not forget so be careful what you promise or lead them to believe. Tape record your first comments to the workforce and revisit these to ensure you're on track with your remarks because the workforce will listen more intently to what you say in your first meeting with them, than any other time.

HELPING GUIDE THE IMAGE YOU WILL ULTIMATELY PORTRAY

Surprisingly, many new plant managers and others who are cast into key leadership roles, approach the job without first ever asking themselves what they want and need to accomplish. Self-examination is an important exercise for the new manager.

Stephen Goodwin, in "Heroes For the Ages", wrote: "The hero wins his reputation through his deeds; the superstar gets his reputation from his image. The sporting values of the Age of Marketing were succinctly expressed in a series of commercials that popularized the slogan 'image is everything.' This is bull, of course. The public knows it's bull and so does the superstar—but what is he to do? The image takes on a life of it's own and things get complicated." (*Golf Magazine*, Dec. 1977, pg. 54)

Goodwin goes on to say: "To become a namesake of an era, he has to be a wonderful player; and beyond that ... he must have qualities that make people cheer for him, admire him, and identify with him." (Page 50)

While business isn't a game, in such a high-exposure position, you will establish an image whether you like it or not. Therefore, what you need to accomplish plays a big part in that image. So, you can sit by and let the inevitable take its course, or you can have a positive influence on your image.

All want to be viewed as highly successful in their job. This is natural. The truth is new plant managers approach their job simply believing if they work hard and put forth their best effort, that particular viewpoint will occur. As Mr. Goodwin notes: "The hero wins his reputation through his deeds." Therefore, employees will remember one's deeds as distinct accomplishments, but it is retrospect in nature. The new leader must look for what helps establish an early image of a potential winner. He or she will want to be viewed as one who has a superb vision and has what it takes to make it happen. Still, any such early impressions must be followed with sound deeds, or the image will sour.

So, what can you do to establish the proper image without overstepping your ability to deliver? This starts by evaluating and, where appropriate, taking some early actions in two areas of concentration:

- Sizing up and correcting key position weaknesses
- Sizing up your operation's global competitiveness

Sizing up and correcting key position weaknesses
No matter what managers think they know about the new operation, one thing they can't adequately comprehend initially is the talent and expertise they will lead. After 30 years in management, I have realized the following: In every organization, there are the doers and achievers, and there are those just hanging on and contributing little, if anything, to the success of the operation.

The workforce knows who these people are. One of the first things the successful manager must do (especially the successful plant manager) is to recognize key leadership positions quickly and if the individuals in those roles are performing satisfactorily. This doesn't always call for replacing people. In fact, using that tactic could be one of the worse things to do. You must create a thoughtful plan of action.

This applies to the salaried and/or management ranks, the place where leadership resides and where management has the responsibility to correct any such problems expediently. I have seen new plant managers who felt a distinct obligation to utilize what they had been given (in other words, to bring along those less productive employees). While admirable, you must balance this action with doing what is best for the operation.

Find the right place for an individual. Many employees are seen as inept because they were cast into a role they were not fully qualified to assume in the first place. The following are two of my strong beliefs regarding this matter:

- Employees are rarely lazy or simply do not care.
- Most employees desire to do a good job and want the company to be successful.

While employees want new plant managers to recognize those inadequate for leadership positions, most do not want such employees fired. Rarely do they openly support such action. The new manager must deal with this disrupting as little as possible and without being perceived as ruthless or uncaring for those cast into roles they were not suited for.

To assess and correct such weaknesses effectively, sit down with those of influence and constructive opinions. Do not use your immediate boss. Your boss's opinion is important, yet most of what he or she knows about your employees came from others or sometimes hasty perceptions (good performers do not always put their best foot forward). So, get accurate information and verify opinions with more than one source.

The most important relationship to focus on is the one between the plant manager and the HR (or personnel) manager. Your mutual respect and relationship should allow for utmost confidence between the two positions. This is critical: Work to ensure you have the right person for this job. Unfortunately, you will initially be forced to deal with the currently assigned individual. Until the level of mutual respect is well established, however, take care in using this person as the single source for the advice and counsel previously mentioned. To accumulate the best advice and counsel, reach out to people who will give you as unbiased an opinion as possible.

Also, take great care regarding your own initial opinions, which may not be the best guide. I once heard a born leader could quickly size people up, that it was an ingrained ability. However, more than once, I have found this to be wrong.

Good performance has unique characteristics but a person's words or reactions can overshadow these. If they are different than what an often high-ranking manager perceives they should be, the tendency will be to suspect the person's capability, especially if the perception was formed casually. This is why I believe no one has a special ability to assess the true potential of others at the offset of any relationship. Further, I question whether a born leader even exists. In reality, leadership is learned, not inherited. Though some people can adapt to a leadership role faster than others, the best honed their skills through solid job experience and, often, excellent mentoring.

You will probably find your selections for such advice and council have been aching to talk. They usually are honored to be asked and often are willing to be candid. Their words will directly reflect how their own subordinates feel about what you've been assigned to lead and more specifically about those who hold key interface manufacturing roles.

After such discussions, share your findings with your boss. This is a courtesy, of course, but it can also reinforce perceptions or lead to a serious conflict of opinions. When the latter happens, reach some mutual understanding about the individual and some collective agreement as to the path to be taken to overcome any perceived shortcomings.

When serious questions or concerns arise regarding a particular individual's abilities, share this with your HR manager for feedback. If you agree the individual is over his or her head then the next step is to correct the situation. Remember that you should make every effort to place such individuals in roles where they can be productive.

Don't forget, most people in key positions didn't ask for their job. Instead, they got there by doing something right. Therefore, management has to accept some of the blame. Since the employee probably displayed substantial skills in getting there in the first place, make every effort possible to retain them in a role where they can perform up to expectations.

Most important, do not let this drag on. Dealing with it early on will say much about your ability to recognize and confront performance problems. It will send the right message to the workforce. However, allowing those who have clear performance problems to continue in their roles will send the message you are either ill-informed or lack the courage to resolve such matters in the best interest of the operation. Further, allowing key position performance issues to continue makes correction later more difficult.

Since people expect assignment changes in the beginning of a new manager's tenure, they are more willing to accept them without major concerns or disruptions. Timing is everything; in this case, it is crucial in setting the right tone.

Here are some major considerations for this most important task:

- In most situations, some in key positions may not be doing the job. For the first step in leading an operation and in setting the proper tone, the new manager must size the situation up and take corrective action quickly.
- Seek some good input before deciding what should be done. Your boss is usually not the best person although he or she may offer some suggestions. One of the best sources are key managers working in another major function, outside manufacturing, although it should be someone you feel will give you an honest and forthright assessment.
- After you receive such input, meet separately with your boss and your HR manager to discuss your findings and your course of action.
- Make every effort to find a place for good performers cast in a role for which they were unqualified. When clear performance problems result from a lack of dedication to job, poor work ethic, etc., quickly provide the individual an opportunity to pursue a career elsewhere.

Sizing up your operation's global competitiveness
In almost every modern business (and definitely within the manufacturing arena), you cannot only look at how well your company stacks up against its competition. Instead, you must take a keen interest in how it compares to the best on a worldwide basis. A company often discovers some more important initiatives it should strongly consider pursuing, from a non-related business and industry. One such initiative would be lean engineering or waste-free manufacturing.

I have seen operations viewed as profitable not having taken even the first step with globally competitive concepts and techniques. This leaves them vulnerable and the new plant manager should present this issue to management and the workforce, especially for one or more of the following conditions:

1. If the operation has a clear history of not meeting promised customer orders. Serious delivery problems and highly dissatisfied customers often exist, with the result being a decline in market share. Though customers dislike what they see, they may be unable to go elsewhere. When this happens, be assured they will only work with the manufacturer until they can find and establish another more dependable supplier.

Sometimes, the workforce isn't aware of the problem. Some may know product schedules or customer complaints have been rising, but others may not. When delivery and customer satisfaction problems do exist, the workforce should collectively assume an obligation to resolve them. Only communications will clarify the problem and the corrective action. If a clear history points to consistently poor delivery, then the workforce should know this along with your expectations and the role they will play.

Two of the worst things a new plant manager can do early on are either to fail to recognize and address issues critical to the success of the operation or to overreact to the unimportant. Do your homework prior to your first workforce communication sessions; and do not delay your first meeting any longer than absolutely necessary since delay can be as bad as being uninformed and ill prepared. Wait no more than one week in fully accomplishing this important task.

2. If the operation has a clear history of poor productivity. Measure productivity against an acceptable industry standard, such as utilizing

Unfavorable Labor Variances to Standard (ULVS), available in most manufacturing operations. Even in smaller organizations, some time standards can be available, such as those estimating labor costs for new products. Still, smaller operations do not always have a formal measurement program aimed at establishing the productivity of their workforce.

Even without a ULVS index, other good indicators exist, such as performance indices directed at universal industry output standards. One such index the automotive industry used was: Total Hours to Produce a Finished Product i.e., a completed vehicle.

3. If the operation has a clear history of <u>excessive</u> scrap, rework, and obsolescence. Excessive is difficult to measure, yet under normal manufacturing conditions, if it exceeds one percent of the total operating budget, you must consider it a problem. Likely, the operation isn't practicing modern manufacturing techniques, which substantially reduce if not entirely eliminate such wastes. Remember that the customer will sometimes dictate inventory levels. In such a case, the company must carry specified inventory, which sets up potentially high scrap and/or obsolescence costs. However, this is done at the customer's demand and, though it has to be viewed as waste, the customer will perceive it as value added.

In truth, the value of added inventory for the customer is usually the fear manufacturing cannot or will not deliver as promised. This, in turn, promotes a Just-In-Case mentality. Numerous manufacturing operations carry added inventory just in case a machine breaks down or in case a supplier doesn't deliver to demand. While it is neither appropriate nor necessary, it unfortunately happens. But when a business is carrying finished goods inventory that grossly exceeds any immediate demand, it is principally due to customers being repeatedly subjected to late deliveries.

Rework is quite another matter and is generally due to poor workmanship, inferior materials, or inadequate procedures. Read *Fast Track to Waste-Free Manufacturing* to learn why rework, in general, and rework stations, in particular, should not be found in a good manufacturing operation.

Instead, focus on resolving the problems, once and for all. An operation is not world class if it requires permanent rework stations because world-class manufacturers resolve such problems, thus eliminating the need for rework as a standard production requirement.

To reinforce the point, I once took over a sizable manufacturing operation and discovered it was riddled with fixed rework stations, which were well stocked and permanently manned. The net result was literally hundreds of thousands of dollars in annual fixed costs. One of my first actions was to demand that the rework stations be torn down and eliminated. I almost had to twist arms before it was accomplished. However, once done, we addressed and resolved the rework problems that had plagued the operation for years. The result was a more efficient and less costly operation, overall. The action I chose to take at the time did pose special difficulties. Though I certainly would not suggest such drastic measures for how to confront every rework issue, it is an example of how a plant manager must sometimes make tough decisions and live with the consequences. I will have more about this as we proceed, but the key is to know when and where such tactics are essential to the operation's success.

4. If the operation has a clear history of excessive downtime. Excessive downtime is another important indicator of an inadequate manufacturing strategy. Manufacturing operations should not be regularly experiencing downtime, a sure sign gross inefficiencies exist. If the operation has planned and budgeted for downtime, you need to address and resolve this fully.

However, before challenging the workforce, you must have a clear history of such problems, not just hearsay or rumor. Get the facts, especially if you will discuss this with the workforce. Look for evidence contributing to downtime over which the workforce has no control; if they exist, let the employees know you are aware of them.

If one or more of the above problems (see Table 2.1) exist, address the matter on the front end. Use the 15-Point Checklist identified in *Fast Track to Waste Free Manufacturing,* which will indicate the seriousness of the problems and to what extent they have spread.

Recapping, these are the two areas of upfront focus, investigation, and potential actions by the manager:

- Sizing up and correcting key position weaknesses
- Sizing up the operation's global competitiveness

Table 2-1. The Need for a Lean Initiative: The Warning Signs

Primary Indicators:	Secondary Indicators:
Failure to meet customer orders	High inventory
Poor productivity	Poor workmanship
Excessive scrap, rework, and obsolescence	High set-up and changeover times
Excessive downtime	Frustrated employees

This should be done, if at all possible after the meeting with your direct staff and before meeting with the extended staff and the production workforce.

THE WARRING ADVENTURE, PART 1

Jim Warring had just been appointed as plant manager of a 20-year-old manufacturing operation producing decorative panels for commercial and privately owned aircraft, as well as parts and components for the service industry.

Before accepting the job, he had learned that the commercial side of the business had some strong competitive pressures, with typically low profit margins. The private and after-market sectors were more lucrative, with high margins but lower overall volume. As a result, slightly over 75 percent of the business was structured to serve the commercial markets, with less than 25 percent geared toward the more profitable private sector.

Even with lower overall margins, senior management clearly viewed the commercial sector as the backbone of the business. Though it would have, of course, enjoyed increasing share in the private and after-market sectors, the company strongly felt the commercial market was where the primary focus should be.

Jim was in his first plant manager job and realized he probably would not influence the company's marketing strategies much. In fact, he'd been told and understood his primary role was to make manufacturing as efficient and competitive as possible.

His boss, Harold Jenkins, vice president of manufacturing operations, went about introducing Jim to his staff of direct reports. Jim only expressed his appreciation for being chosen and noted he looked forward to working with them. Toward the end of the meeting, Phil Tanner, the HR manager, asked if he wanted him to schedule some meetings with the general workforce, but Jim politely deferred, saying he would do that later after he was more organized.

In the ensuing days, Jim learned more about his operation. With Harold's help, he met with two of his boss's peers, both of whom were key executive managers and headed up other major functions.

Jim's first meeting was with Fred Johnson, the company's sales and marketing manager. After some introductions, Fred opened the conversation. "Harold didn't tell me much, other than you wanted to meet with me to get some input."

"Yes, that's correct," Jim replied. "Let me start by saying how much I appreciate you taking the time and I want to assure you it's important. I sincerely need some of your thoughts and opinions about our side of the business. That's if you don't mind."

"Well, it's a refreshing change," Fred retorted. "What I mean is having someone in manufacturing ask us our opinions," he clarified. After a short pause, Fred continued. "So, how can I help you?"

"Being new here, I need to understand where we have shortcomings. I want to shore them up if I can and I was hoping you could help. Though I'm not out to clean house, I do want to address this issue head on and correct any weaknesses."

"Well, since you've asked, I'll be as candid as I can with you," replied Fred. "First, on the positive side, Bob Frisman is the best man you have."

"I've heard the name. What can you tell me about him?" Jim replied.

"He's your senior manufacturing engineer over the private and service sector, P&S as we call it. Anyway, he's a great guy, has his act together, and is easy to work with. In fact, if I were to take a poll who most of my people think is the best talent in manufacturing, I'd say he'd win hands down."

"Really?" Jim asked. "Well, I'm a little confused. If he's that good, why isn't he working on the commercial side?"

A smile spread across Fred's face. "We've been waiting for someone in manufacturing to recognize that. I was recently chatting with one of my senior people, and he was saying the same thing. But, if you're thinking about putting Bob in Commercial, you're going to have a real fight on your hands, especially with Harold."

"Harold? My boss?"

"That's right. He was instrumental in placing Bob in P&S after getting a lot of flak about delivery problems. He and Ken Watkins, the man you replaced, felt the problem was a lack of good manufacturing engineering (ME) talent in the area. Obviously, they were right because Bob was key to helping get the operation back on schedule."

"How long has he been there?" Jim asked.

"Oh, since he started. Two years?"

Jim reflected, mentally placing a stamp on the subject for later consideration, then proceeded on. "I'll definitely keep that in mind, but please continue,"

"Hmm," replied Fred, thinking carefully before commenting further.

"Obviously, I'm not in a position to know how well all your people perform, so anything I have to say would be based to a large degree on what I hear from my people."

"Fair enough," replied Jim.

"This is a little awkward, but I'd have to say that you need to look at Cindy Mabrey and Chuck Henson. Cindy heads up customer interface with Aklin Industries, one of our larger commercial customers, and Chuck is the foreman over the Aklin line. Aklin isn't happy with us, and rumor is they're looking for another supplier."

"What's the problem?"

"Got an hour or two?" replied Fred, chuckling. "It's like this. Cindy and Chuck have been around a long time. They held the same positions when the line they're managing today was making parts for Polen. When Polen bailed out for another supplier, they made it clear they were highly dissatisfied with Cindy and Chuck. Of course, Cindy and Chuck took most of the heat for what went wrong and much of Polen's dissatisfaction was because the department was behind schedule for so long. But another problem was that they got little satisfaction out of Cindy when they called her about it."

"I understand. So, what's your people's opinion of them?"

"Frankly, Chuck wouldn't rate high. They think he's extremely negative, and he has this way of making it clear he doesn't want them in or around his turf. On the other hand, I've been here long enough to remember working with Chuck on a number of projects and, as I've told my people, I really didn't find him to be that way. But, you know, things change and not always for the best." Fred related much of the feedback he had heard. Much like Chuck, Cindy was viewed as being uncooperative and at least one of Fred's key people had openly expressed she should be fired. He also covered a couple of other key positions he felt Jim should look into, but Fred didn't appear as concerned about them. Outside of that, he made it clear he and his people had no particular qualms and, in fact, felt many good employees were working in manufacturing.

Jim thanked Fred for taking the time and let him know how grateful he was for the feedback. He ended by promising he would treat it in a confidential manner. Fred, in turn, let Jim know he appreciated he was interested in his opinions and that he did not worry about Jim misusing or misrepresenting any of the provided information.

Later, Jim met with Justin Pearl, the company's chief financial manager. He proceeded in the same manner he had with Fred Johnson and received similar feedback about Chuck and Cindy. However, when Jim came to Bob Frisman, he got a totally different viewpoint. Justin admitted he didn't know much about Frisman's engineering talent, but he sure knew how terrible he was with budgets and forecasts. Bob was responsible for preparing monthly forecasts for engineering expenditures and for forecasting scrap and rework costs.

"He does a terrible job of it," remarked Justin. "I stay in hot water with the powers above trying to explain why we can't do a decent job of forecasting expenses. I can't tell them it's Bob's fault. They'd tell me correct it or see he's replaced. Frankly, Jim, I've been over this before with your predecessor and he did nothing about it."

Jim interrupted. "I've obviously hit a sore spot. You're upset about this and I can understand why. I will definitely look into it and see if I can get it corrected."

"I hope so," replied Justin.

Jim decided not to press the issue, and thanked Justin for his time. That afternoon, Jim met with Harold to give him the highlights of what he had heard. The feedback didn't surprise Harold, though he didn't fully agree with either assessment. While Harold agreed Cindy posed a problem and it was something Jim needed to address, he felt her knowledge and general abilities were an asset to manufacturing. He adamantly disagreed about Chuck and made it clear he wouldn't support any effort to move him out of P&S.

"Frankly, and with no disrespect intended, they have their agenda and we have ours," remarked Harold. "Chuck is absolutely instrumental to our operation. They simply don't have all the facts. And, with regard to Bob, he's where he is for a special reason and I won't support any effort to take him out of P&S."

"Well, that's the main reason I wanted to let you know about this and get your feelings. Still, right or wrong, there are strong negative perceptions that shouldn't be ignored."

"Oh yes, I've heard it before. But, in all fairness to you and what you want to achieve with all this, let me give you a slightly different perspective," responded Harold.

Jim listened as Harold related a lengthy story behind the delivery problems with Aklin Corporation and other customers within P&S. He saw the problem as a lack of leadership at the plant manager level. He insisted that Ken didn't demand enough of his people and focused too much attention on expenses rather than customer satisfaction, which is why Jim was now there. According to Harold, Ken consistently avoided overtime in order to keep expenses in line with budgets and forecasts, which was fine but was absolutely the wrong thing to do when customers' demands were not being met.

Harold paused, and Jim decided to test the water with his own strong conviction, learned from his previous experience with the application of lean manufacturing principles, concepts, and techniques. "Wouldn't you agree the answer to that problem isn't overtime but rather resolving the problems that make it necessary?"

"I absolutely agree," Harold said. "However, until the problems are fully addressed and resolved, overtime is necessary and that's what you do. Do we agree on that?"

"Definitely," said Jim. "But, I'm having a difficult time understanding how what you've covered applies to what Fred and Justin had said. If you listen to them, they are talking about some people with an inability to do the job or some apparent lack of desire to work effectively with peers and customers."

"Well, that could be true, and I'm not going to sit here and tell you there may not be some training that's needed. All I'm saying is Cindy and Chuck were between a rock and hard spot because they were doing what they were told to do. They were being good troopers in the process by not blaming their actions on someone else."

"I think I see where you're coming from," replied Jim.

Harold went on to point out why Bob should stay in P&S. Essentially, it boiled down to a strong feeling Bob was making tremendous progress in resolving problems plaguing the area for years. Clearly, Harold did not want to see the effort decline in the least.

Jim explained to Harold he didn't have any immediate notions about moving Bob but he wanted to leave his options open. Harold assured Jim, in turn, he would let him run the plant the way he felt he should. But, clearly, some restrictions had been placed on just how far Jim's freedom went when it came to certain people.

"All I'm really interested in," Harold continued, "is having an operation that meets its stated goals and objectives and where people are treated fairly in the process. I'm fully confident that you'll make it happen."

Later that day, Jim sat in his office scribbling notes while the conversations he held earlier were still fresh in his mind. His key findings were that two high-level managers had agreed a few key manufacturing positions had problems and needed to be addressed. Other than that, nothing negative had surfaced, other than one person who had some difficulty forecasting and budgeting expenses. Harold, his boss, had essentially agreed some problems existed but felt these were extenuating circumstances over which the individuals in question had no control.

In addition, Jim had heard and seen enough to know the factory had a bigger problem, which was the lack of world-class manufacturing thinking. This prevented the application of concepts, practices, and techniques that would resolve a history of delivery problems and dissatisfied customers. One thing for certain, the key positions in question weren't responsible for that.

Before heading home, he toured the plant. The second shift had just started and as he walked through the entire factory, stopping occasionally to say hello and to quiz people about their job, he made some mental notes along the way.

First, he saw too many inactive people. Although Jim didn't expect to see everyone slaving away, he decided to use a technique he had learned in his last job. It involved counting people at their workstations, performing productive work. This technique worked better with two people, one of them counting those performing productive work and the other counting everyone else. But Jim had learned another way to get a fair overall estimate. He would isolate an area and note who was present at their workstations. Then, by looking at daily attendance records, he got a decent estimate as to how many people were present. Jim had learned that in less efficient operations, it was common to find fewer than 25 percent of the workforce performing productive work, but he was surprised when his number was only 16 percent. This didn't prove anything yet, other than at that given moment, 84 percent of the second shift wasn't performing value-added work, but it implied a real problem.

Jim also noted all the inventory. He found it on the floor, in storage racks, around machines and equipment. A swipe of the finger on one set of boxes supplied enough dust to tell him it had been there for awhile. His past experience quickly told him he was observing a substantial problem as well as a golden opportunity.

The following morning he met with Phil Tanner, the H.R. manager, to discuss yesterday's feedback and to decide when to meet the workforce.

"Phil, it's extremely important to have accurate information and to take action in certain areas, if required, before meeting the workforce. That's the reason I've delayed the meetings up to now. I'm fully aware we can't hold off forever, but I need to share some feedback I got yesterday and get your thoughts. Then we have to decide what actions, if any, we need to take and finally the timing and location of the meetings."

"Fine," replied Phil, as he pulled out a notepad and pen in a silent show of readiness.

"First of all, I met with some managers outside manufacturing to discuss key manufacturing positions and to get their thoughts on the matter. I'd like to keep the manager's names confidential, for the moment, and focus on what they had to say," Jim began. Phil nodded his agreement, although he was a little puzzled.

"Two names surfaced repeatedly: Cindy Mabrey and Chuck Henson. Before I get specific about what they had to say, is there anything you can share with me about them?"

"Well, their names have come up before. They've had a history of problems working with fellow associates and customers."

"That's precisely the feedback I got. But, if everyone's known about it, why hasn't anyone done something about it?"

"Oh, they've been given verbal and written warnings."

"Tell me more," requested Jim.

"It's like this. They've been here a long time, and their peers in manufacturing like them, especially the hourly ranks. They're seen as this wall between the department and the outside world, who they think don't understand their problems or don't care. Chuck and Cindy are talented. No question about that. It's just this matter of how they're perceived by those outside the group and, more important, by customers."

"Has anyone thought of placing them in other, more effective roles?"

"Not really, although, there has been discussion about firing them."

"And?"

"And, it always ends with them being seen as essential to keeping the department going."

"Phil, I want your opinion."

"I hate to say this, but if it were me, I'd replace them. Sometimes you have to bite the bullet and do what's best for the company. Don't take me wrong. I like Cindy and Chuck and would hate to see them leave."

"Again, I'm surprised they haven't been considered for some other role," said Jim.

"Well, the rule around here is you either cut it in your job or hit the door," Phil replied.

"Really? Well how do you feel about that?" asked Jim.

"I'm not sure how to put this, so I guess it's best just to say it. That philosophy, so to speak, comes from the top. From Harold. He's had quite an influence on how daily manufacturing activities are conducted and who's selected to do what."

"Rightly so," replied Jim. "After all, he is VP of manufacturing. Of course, if the facts were known maybe he feels he has had to get more involved than he had any desire to."

"True but, right or wrong, that's the way it's been."

"Fair enough. Let's move on," Jim replied. "I would like you to talk to the necessary people, and I would like your recommendations tomorrow as to where Cindy and Chuck could possibly fit outside of their current roles. In addition, I want your thoughts about who could replace them. In the meantime, I'll have a discussion about the matter with their boss, Joe Thompson. What do you think Joe's reaction will be?"

"No question. He'll fight tooth and nail to keep them where they are," replied Phil.

"We'll see."

Jim and Phil then went on to discuss Bob Frisman. Phil was as high on Bob as everyone else, but he added something new to the equation. Recently, Bob had come to Phil, at the suggestion of his manager, Zack Milen, to discuss his frustrations. Simply put, he wanted out of P&S to where he could test his engineering talent and ability. He felt he was just a general foreman and that he was doing little, if any, manufacturing engineering work.

"How do you think he would feel about a role in commercial?"

"I think what he's doing is more important than where he does it."

"If that's the case, resolving the matter becomes a little easier," Jim replied. "I'd like to meet Bob myself after discussing it with Zack. I want to get to know him better and understand how he sees things."

"I think that would be great," replied Phil.

Later that day, Jim met with Zack, who was fully supportive of Jim meeting one-on-one with Bob. Jim then proceeded to meet with Joe Thompson, the operation's Production Manager, who held a crucial position, in Jim's judgment. The feedback he received from Harold about Joe was that he was a good trooper, had excellent product and process knowledge, and was respected by his peers and associates.

Jim began, as he had with Phil, letting Joe know he was there because of some feedback he had asked for regarding key positions. Jim said he wanted to share some of this with him and get his thoughts. As he was finalizing his feedback about Chuck and Cindy, he noticed that Joe became visibly irritated. When he was finished, he waited for a response.

"I don't know what to say," began Joe. "I really wasn't expecting this. Chuck and Cindy are tops in my mind and critical to keeping their section running effectively." He paused a moment. "If I could be frank, I'd have to say there are more pressing issues."

"Don't get me wrong, Joe. There are other important issues, but this isn't one we can ignore."

"Well, I have to give you a benefit of doubt, I guess. So, how can I help?"

"Well, first, I'm not out to punish anyone. Based on everything I hear, Chuck and Cindy have a good deal to offer in terms of their knowledge and ability. But I also know something has to be done to correct a negative impression they have. We can't have people in key roles who have difficulty interfacing with peers and associates in other functions and, more important, who are irritating customers. However, if the impression I've been given is wrong, now is the time for you to speak up."

Joe agreed Chuck and Cindy had a reputation for being difficult to work with but felt this was overblown. Joe contended the root of the problem was manufacturing processes engineered incorrectly, didn't work properly, and thus created an inability on the department's part to meet customer requirements. As a result, Chuck and Cindy were placed squarely between those who were responsible for the design and development of those inadequate processes and customers who applied unrelenting pressure to deliver as promised. Further, he noted, they were restricted in how to overcome the problems because of management dictates about keeping overtime to a minimum while fully meeting established budgets and financial forecasts.

Jim listened patiently, only occasionally interrupting for clarification of certain points. He then assured Joe he had taken note of the special circumstances and promised to get back with him. Jim changed the subject and asked Joe for his thoughts about Bob Frisman.

Joe reinforced everything Jim had heard about Bob and mentioned he had given him duties and responsibilities, which were not manufacturing engineering oriented. When Joe was forced to let the general foreman go, he asked Bob to assume a leadership role until he could find a replacement. Bob had displayed a great deal of gusto in the assignment and had shown he could get things moving in the right direction. Thus, Joe admitted, he had been slow in hiring a new general foreman.

"How long has he been in this role?" asked Jim.

"About three months," replied Bob.

"While your intentions were good, I'm sure you are aware that Bob isn't exactly a happy trooper. Phil Tanner tells me he is asking for a job somewhere else."

"No. I wasn't aware," replied Joe, surprised with the news.

"You weren't?" asked Jim. "Well, that's a problem in itself and something we'll clearly have to address. But I'm going to ask you to keep that between us for now."

"All right."

Jim went back to his office to take care of other pressing duties. He wanted to think about and recap everything he had heard before meeting with Bob Frisman. Jim had learned from a former boss to take time each day to keep a running business diary. The key was to keep it handy at all times and review it on a frequent basis. When an item was fully resolved or no longer an issue, he removed it from the diary to insure he had a running record of current issues. That kept the diary smaller and assured him he wasn't spending a lot of time wallowing in the past. These are his current notes:

Meeting with Phil:
  Gave Chuck and Cindy verbal/written warnings
  Agrees it's a problem
  Mentioned Bob had met with him
  Bob doesn't feel good about current role

Meeting with Joe:
  Widely different views/reasons than Phil
  Spoke of some serious process problems in the area
  Feels this has had an influence on the bad impression
  Agreed he has used Bob more as a foreman than Mging. Eng.

Joe and Phil: apparent lack of communication?
Joe unaware of Phil's meeting with Bob

To Do:
  Find out if Bob is interested in Unit Mgr. role
  Staff level communications?
  Get workforce meetings clearly established/communicated
  Cover all this with Harold later

In making the notes, Jim was recapping everything in his mind. Evidently, most of those he met shared a negative perception of Cindy and Chuck. Though he had heard enough to believe Chuck and Cindy had something to offer, it clearly wasn't in their current assignment. They were having a bad influence on other functions and on customers, who were extremely dissatisfied with the relationship. Although Joe highlighted process problems, over which Chuck and Cindy had no control and had presumed certain management restrictions for overtime and expense control, Jim couldn't accept this as an excuse. If this were indeed the case, then Joe wasn't doing his job by allowing the matter to go unresolved. That was, of course, unless he had been restrained by someone above. Jim needed to know more.

He was concerned that Phil had held a meeting with Bob and hadn't let Joe know about it. If true, then he had bigger problems to address, which, of course, was the matter of appropriate communications between the staff and, perhaps, throughout the ranks.

Jim felt Bob could be missing his true calling. He wanted to listen to Bob and see if he had any interest in changing his career focus toward unit management, perhaps in Commercial. Bob had definite leadership potential, but he could leave out of sheer frustration if some clearer direction wasn't established regarding his career path.

Last, but not least, time was running short. Jim had been on the job for three days. He knew he had to meet with the workforce and had to bring something of value to the meeting. After all, stopping production and holding communication meetings was expensive. Therefore, if possible, he felt such meetings should be value-added in nature.

"Bob, it's nice to meet you," said Jim, after he had introduced himself. "I've heard great things about you and what you've done to help P&S resolve some serious problems."

"I suppose I've done my part," he replied, "but plenty of problems exist. I wouldn't want to mislead you on that."

"I'd like to come back to that a little later, but for the moment I want to focus on you with special regard to your aspirations. You see, you're here because I've heard enough to know you have a lot of talent and ability. I've also heard you have some misgivings about your current role and have requested a transfer."

"I've discussed the matter with Phil, if that's what you mean," said Bob.

"Yes, I've heard. Would you give me a brief overview of what it was all about?"

He gathered his thoughts. "It's like this. When I came aboard, I thought I was getting with an organization where my education would be fully utilized. I have a degree in manufacturing engineering, and I think I'm better than average. But, from the start, I've been just a glorified foreman. I was willing to help Joe out, but he told me it would only be until he could find a replacement for Ben Womack, who he let go shortly after I arrived. It's been three months, and I just don't see any end in sight."

"I see," said Jim. "Have you talked with Joe about this?"

"More than once," Bob replied.

"And?"

"He keeps saying as soon as he hires a person. Says it's been tough finding someone."

"He's probably right. Finding someone who's qualified for supervision isn't easy. Many people see it as a dead end, and the job is difficult. It takes a special kind of person."

"True. I've done it long enough to understand what you're saying," responded Bob.

"Forget for a moment how and why you got there. How do you feel about this type of work?"

"I rather enjoy it, and my ME background helps dealing with equipment and facilities. The people on the floor respect someone who is mechanically inclined and understands how to keep the equipment running properly instead of always having to call the maintenance department. If I've done anything constructive, I've reduced downtime. Many times, I'll service or fix the machines rather than wait for maintenance. They take forever to respond sometimes."

Jim listened patiently as Bob again paused briefly before continuing.

"But, given all that, my desire is to return to some real ME work."

"I understand," said Jim. "But could it be you might have an interest in supervision if in the right kind of position, say as a unit manager?"

"A unit manager position!" Bob retorted, obviously surprised by the comment.

"As I see it, the unit manager slot is an extremely important position and, whether it's presently viewed that way or not, I intend to raise it to that level. I also believe unit managers should have an engineering background, for the reasons you mentioned. I'm not saying they should spend precious time repairing equipment because that's the maintenance department's job. But knowing what makes the equipment tick helps them to resolve minor problems. That way they don't take a lot of nonsense from anyone."

Bob smiled broadly, enjoying what he was hearing. Before the conversation concluded, he assured Jim he would give becoming a unit manager some serious consideration, if it were indeed what Jim felt would be good for him and the operation.

They discussed other problems, most relating to original equipment design and installation. It was just another of the growing examples Jim was coming to recognize, of manufacturing processes throughout the operation, that didn't work as intended.

Later that afternoon, Jim met again with Phil Tanner. He brought him up to date on his findings and then asked him about a troubling matter. "Phil, regarding the conversation you had with Bob; Joe tells me he knew nothing about it."

"He certainly did," Phil replied, his voice raising a bit. "I told him about it shortly after Bob came to me."

"What did you tell him, specifically?"

"Nothing specific other than Bob had asked to see me to discuss his career."

"You told him nothing about Bob being frustrated?"

"I consider that a confidential matter between an employee and the HR manager. If I related everything I was told, before long I wouldn't be trusted by anyone."

With a sudden firmness, Jim interrupted. "You're not to keep secrets from fellow staff managers, regarding what their employees say to you, unless, of course, it has to do with unethical behavior. In that case, I expect you to come straight to me. Second, put yourself in Joe's shoes. It was embarrassing for him. Plus, you didn't let me know you hadn't discussed it with him. Finally, if people view you as someone they can approach to unload every concern then that also has to stop."

"I'm sorry," replied Phil. "It's obvious you feel strongly about your convictions."

"In all fairness to you, let me explain. A manufacturing operation is going to succeed if it has a management team which works well together, has the same basic convictions about what is best for the operation, and has a high degree of respect for each other. The firm with a suspicious management team that doesn't like or respect one another is going to fail. It has to be like a good football team. We don't have to agree with everyone else on everything, but when it comes to performance, absolutely everyone on the team has to be devoted to the same objectives."

"Well put," Phil noted. "And you can depend on me to keep it in mind. However, let me tell you why I took that position."

"It isn't necessary unless you have to get it off your chest. I honestly don't care why, only that we start from here and plot an appropriate course for the future."

"All right," Phil replied. He seemed disappointed Jim discouraged further comment.

They discussed possible reassignments for Cindy and Chuck and qualified replacements for them. Phil had similar positions for them in P&S but drew a blank on possible replacements. Jim let him know that wasn't good enough and suggested he think about it again and get back with him first thing in the morning. He ended the conversation by reminding Phil they had to finalize plans for workforce meetings.

Jim was glad that a tough day was ending. He needed some rest after spending most of the previous night planning how to handle today's matters. It hadn't turned out exactly as he had anticipated.

On the way home, he mulled things over. He felt bad about almost losing it with Phil, but his response to what appeared to be thoughtlessness and disrespect for fellow peers had appalled him. On the other hand, he liked many things about Phil. He was intelligent and a quick read of others, which was definitely an asset. Jim also felt comfortable with Joe but wasn't sure he could keep his personal feelings out of the way of sound professional judgment. Time would tell. Both had disappointed him a bit. Jim felt as if he were playing truth or consequences with them. He worked too hard to pull things out, and neither man had been totally open with him regarding Phil's meeting with Bob.

Maybe it was just part of being a plant manager, he thought. After all, in this new role, he was having his first experience interacting with others in such a position. He had much to learn, and more surprises were sure to come. As a result, he had another restless night.

The next morning, he planned his day on the way to work. Right after the business meeting, he would meet with Phil and settle the Chuck, Cindy, and Bob issues. Then he would finalize plans for the upcoming communication meetings and spend the rest of the day preparing for them. One additional chore was meeting with Harold to bring him up to date and get his blessing on his final plans. But it didn't go that way.

Upon entering his office, Harold called. Phillip Brooks, president of the operation, had called a special meeting. Harold said he would explain everything on their way downtown to Brooks' office, a good 20-minute drive. Jim called Harold's secretary back to confirm Harold would be driving. When she said yes, he asked if she knew anything about the meeting. Her only reply was, "The Aklin account."

On his way over, Jim was thinking. "We must be in trouble. They must have called Brooks and complained. He's probably going to lay down the law. Either get things straightened out or he would be forced to find someone who could." As it turned out, the news was even worse.

"Brooks received a formal letter from Aklin yesterday stating they intend to move their business elsewhere at the end of the current contract, which will expire at year end," explained Harold. "That's less than five months away," he continued. "Legally, there's a question of appropriate notification, per our contract with them, which states a minimum of six months notification, but whose gonna squabble about one month? Not Brooks, I'm sure."

"Did they say why?" asked Jim.

"Poor quality and substantial delivery problems. On at least two occasions, we've shut them down. That was before your time, of course, but you have to deal with it now."

"I understand, but what, if anything, should I be prepared to discuss?"

"Nothing really. Brooks won't expect you to have all the facts this early. I'd say listen and take notes. If I were to predict his position, I'd guess he's going to expect us to visit Aklin and see if we can salvage the business. But he can be unpredictable."

Unbeknown to them at the time, they were in for another surprise.

Upon entering his office, they not only found Phillip Brooks but also Alan Cummings, vice president of operations for Aklin. He sat at the conference table with Brooks and Fred Johnson, the sales and marketing manager Jim had met a few days before. After introductions, Brooks explained Alan was there to visit with Fred and they had dropped by his office to say hello. Brooks inadvertently mentioned Jim and Harold were on their way to meet with him and Cummings asked if he could join them to hear manufacturing's story regarding the situation. Brooks, of course, said yes.

Harold's face showed he didn't like the developing situation. He could do little about it now, so he welcomed Cummings and opened by pointing out that with Jim now aboard he was confident Aklin would see significant quality and delivery improvement.

"Oh?" replied Cummings. "So, what's your plan, Jim?"

"My plan? Frankly, I'm still getting my feet on the ground."

"I know, but what's your basic philosophy about manufacturing. In other words, what will you bring to the party that's going to make a difference?" Cummings challenged.

Jim realized he was unexpectedly facing one of the more crucial challenges of his career. He decided not to panic and thought carefully before responding.

"I bring a philosophy that's strongly aligned with world-class manufacturing principles. Products can always be delivered on time and with outstanding quality if a manufacturer places the right kind of emphasis on continuous improvement and fully utilizing the proper manufacturing concepts and techniques. But equally important is having the right people in key positions, who not only understand those principles and techniques, but feel a strong obligation to put them into practice. I'm now actively involved with analyzing the issue of key positions and will be making some announcements soon."

Jim continued. "Having said that, I'm sure it's little more than just words to you, but if Aklin gives us another chance, I promise you'll see substantial improvement."

Jim glanced at Harold, who smiled warmly. He liked what Jim had to say.

Cummings was quick to respond. "Well, I can assure you we have no desire to take our business elsewhere. Setting up a new supplier requires extra work we'd rather avoid. But we've had promises from you before and we can't afford to continue on this path."

"I understand," replied Jim. "If you give us another chance, you won't be disappointed."

"Tell you what," Cummings remarked after a moment. "If I can set it up, would you be willing to come to our manufacturing facility in Camden and put on a presentation for the team regarding your plans to get things back in order?" Before Jim could reply, he turned to Harold. "Harold will probably want to be there also."

Harold nodded in agreement as Jim immediately replied he would be honored at the opportunity. Cummings said he couldn't make any promises this would change the decision, but he felt it wouldn't hurt.

Roughly one hour later, everyone departed with the understanding Cummings would get back to Johnson on a meeting time and date if he were successful in setting one up.

Harold let Jim know he had handled things extremely well. Jim tried to take the opportunity to discuss key positions and was surprised when Harold remarked, "Jim, do what you think is right. Just keep me informed. You have my support, so decide what's best for the operation. As I've told you, the most important thing is you treat people fairly in the process. Good enough?"

"Good enough," replied Jim, relaxing for a moment. He was pleased with how he had conducted himself and with the confidence Harold had apparently placed in him.

Jim was about to learn that self-satisfaction as a plant manager is usually short lived. On returning to the factory, he was forced to handle a number of relatively serious problems relating to equipment downtime, along with material shortages. By the time he worked through all this, it was late in the day and he still hadn't met with Phil. He asked Phil if he would stay a little late, so they could continue yesterday's conversation. Phil agreed.

Jim knew exactly what he was going to do. What he didn't know at the time was that he was about to make his first big mistake.

HOW AND WHEN TO LISTEN CAREFULLY

Listening carefully, as it applies to the business world, means listening to what people are saying. Most important is to assure you never make assumptions about what people think, feel, or expect as a result of any single comment they may make.

Whether a plant manager prefers it, most people they direct will not be totally candid. It's just a fact. A polite, yet probing style is usually warranted. This not only applies to a plant manager, but to anyone in charge of being the ultimate play-caller for a business.

How to listen carefully is one matter. When to listen carefully is quite another. Being human, we have distractions, disruptions, and frustrations interrupting our undivided attention. Therefore, understand when you must provide an atmosphere where listening carefully can best take place.

A friend once told me: "Never assume you understand precisely how people feel until you have asked them at least twice. You're better off irritating people with repeated questions than assuming you know their answer, based on a single comment. With important issues, you must know precisely how they feel. The only way you'll know for certain is to ask more than once."

That philosophy doesn't apply to every conversation. The need is directly related to the level of importance.

On occasion, what I thought someone had committed to was entirely different. Often, what I heard was what I wanted to hear, not what they were telling me. Though I cannot advise you how to know people's inner feelings, you must give them the benefit of doubt. My friend aptly said: "The only way you'll ever know for sure is to ask more than once."

You can ask in many ways, without being redundant and, therefore, potentially irritating. Do not run to your boss (or others) about specific instructions or directions, just to ensure you got it right the first time. That's a sure way to gain the wrong kind of reputation.

No matter whom you are dealing with, first listen carefully to yourself with a method I call the strainer process. In this doing this, you are cleansing your thoughts of impurities and uncertainties. This strainer process calls for numerous actions on your part, prior to you finally communicating and implementing something. In order, these are:

1. **Examine the idea yourself.** Take time to consider other people's feelings about the change you are about to make. Put yourself in their shoes and consider how you might feel. Use self-criticism and imagine a worst-case scenario. Consider everyone important and imagine his or her reaction. Always include your boss, your direct staff, your peers and, then, all others including the union leadership and community officials, if required.

2. **Consider other alternatives.** Stop, take a deep breath, and think about other alternatives. Come up with ideas you haven't yet considered. Be totally flexible in your thinking. Think of what you would do if you owned the company, were in total control, and could do anything you like. Most important, forget the usual restrictions placed on your thinking. If you have been thorough and can't find a better solution, then forget about the other alternatives and focus on how best to implement your original idea.

3. **Seek out the advice of one you respect.** Bring people in who have no bridges to burn or any preconceived notions about what is best. In most cases this will be past associates who are not working for your company. Tell them you need advice. They will appreciate your confidence in them as you tell them the issue and how you plan to address and/or resolve it. You should include enough details so they understand the big picture. You might be surprised how clearly they see answers or how quickly they produce solutions you haven't considered. If no such a person exists, pick someone within the company, preferably from an entirely different function outside manufacturing.

4. **Update your boss before pulling the trigger.** In each case, any significant change to the operation should be covered with your boss

prior to finalization and implementation, especially if the change fits within one or more of the above three categories.

Listening carefully, as applied to business decisions, means more than just listening intently to a person's words. It means taking serious time to question and listen to yourself. Follow this by listening to the thoughts and ideas of peers and associates, people you respect and, in almost every case, your boss.

Jim spent time bringing Phil up to date on his meeting with Brooks and Alan Cummings. The adrenaline was still coursing through his veins as he proudly explained he had been given an opening, took it, and, as a result, would visit with Aklin's management team in the Camden facility, to salvage the business. Phil was obviously impressed.

Jim noted he had also spoken with Harold about Bob, Chuck, and Cindy. Harold made it clear it was Jim's decision to make and he could depend on his support. Phil seemed surprised, but didn't press the point.

"Now," said Jim, "I think I've made my decision, but I want to run it by you first."

"Fine," Phil replied. "Let's hear it."

"I met with Bob and I'm convinced, along with everyone else, that he's an outstanding talent. I asked him to consider a unit manager's role, and he seemed interested. We have a significant problem but a grand opportunity regarding Aklin. If I can convince them to give us one more chance and then implement changes making an immediate difference, we will probably keep their business. I could see Cumming's eyes light up when I mentioned changing key assignments. So, I want you to offer Bob the unit manager position for the Aklin account. Since you didn't see another adequate role for Cindy or Chuck in another area, they will continue but under Bob's direction. Over time, he'll monitor their performance and make some decisions as to the best course to take. The biggest loophole is who will replace Bob in P&S."

Jim paused, waiting for Phil's reply. Phil leaned back in his chair and cleared his throat.

"Well. We can do that, however, I've been working on this matter since we last discussed it and I've found out a few things you should know."

"Of course, please do," encouraged Jim.

"Bob came to me again, just this morning, to talk about things. I let Joe know about it and covered everything Bob had to say with him."

"Good," responded Jim, "but about the meeting?"

"Bob is extremely frustrated. He mentioned his meeting with you and what you had to say about becoming a unit manager. He is dead set against it."

Jim was utterly surprised, but again encouraged Phil to continue.

"He's worried about being a glorified foreman and never really utilizing his ME skills."

Jim interrupted. "Phil, that's the same conversation I had with him. I tried to tell him he could still use those skills. When we were together, he seemed to understand and was supportive of the role I described. Maybe he needs some gentle persuasion from you. Why don't you meet with him again, let him know I specifically requested he take the position and also that it could be the best thing for his career?"

Phil thought and replied, "All right, I'll do it. I'll get on the paperwork for this right away. I'm going to need your signature on the job offer form, however."

"Get it to me and I'll sign it," said Jim.

With that, the conversation continued with a discussion about a number of candidates that Joe should interview as potential replacements for Bob. They also agreed on a date when Jim would conduct his meetings with the workforce. It was two days away. They finished by agreeing that Jim should be the one to break the news to Joe about Bob.

After a long day, Jim was ready to wrap things up. He decided it might be appropriate to celebrate just a bit. There hadn't been much of that since he had taken on the assignment. So, he called his wife, Ginger, and they had dinner out.

The next morning, the phone rang. It was Harold's secretary. "Jim," she said, "could you come to Harold's office. He needs to talk with you."

"When?"

"Right away, if possible."

"Sure," he replied.

When Jim entered Harold's office he could see he wasn't a happy trooper. Harold had a horrible frown on his face as he glared at a piece of paper he held. After Jim had taken a seat, Harold raised the document and then gently tossed it across the desk toward him.

"What's this?" he asked. His words slow and separated and his voice firm.

"I don't know," replied Jim. "Let me take a look."

He intended to review it carefully, but it wasn't necessary. It was a job offer form and the candidate outlined in heavy black printing was none other than Bob D. Frisman. Jim hardly knew where to start. First was the surprise of Harold getting the document before he had a chance to review it. Though he thought Harold had said such decisions were his, it was apparent he was more than just a little upset about the matter.

"I hardly know what to say, Harold."

"I can understand that," he replied.

"You said yesterday on our way back this was up to me. You specifically said you would support me."

"We obviously have a communication problem," remarked Harold bitterly. He waited before continuing, and Jim could see his expression soften.

"Look, Jim. I'm upset and sometimes I let my anger get the best of me. I suppose what we really need to do is talk this through and understand each other. To start, I was shocked when I received this, this morning. You obviously had no idea it came to me, but policy says I review and approve all job offer forms for any salary grade above 40, which unit manager positions fit. I thought Bob was going to stay in the P&S. Apparently, you thought yesterday's comment meant you could include him in the assignment changes."

"That's definitely the case," replied Jim. "I would never intentionally go around you on such a matter."

Harold nodded his head in agreement but didn't reply. They sat for a moment waiting on the other before Jim decided to pursue the conversation.

"Where do we go from here, boss?"

"Well, one thing for sure, you can't move Bob out of P&S."

"If you don't mind, would you tell me why?" asked Jim.

Harold sighed, signaling his disappointment in Jim's question. "Bob is there because I put him there. I won't sit here and criticize your predecessor, but I'll just say I had to step in and make this move. In doing so, I told one of our biggest customers, Warren and Associates, that the key to getting and keeping P&S back on schedule was going to be Bob Frisman. That's happened and there has been no end to the praise Warren and others have given Bob. I simply couldn't allow him to transfer, even if I wanted."

"I see," replied Jim, "Sorry. I didn't know the whole story. Then you need to know he may leave on his own at any time. He is a severely frustrated employee."

"Oh?"

Jim related his conversation with Bob, as well as the feedback Phil had given him. "The point is, he may be leaving P&S, one way or the other."

"I'm sure you and Phil can correct that with a little persuasion," tested Harold.

"That could be if you think we should. But, it still leaves me with quite a mess on my hands. I only hope Phil didn't see it as a done deal and hasn't already approached Bob."

Harold asked that he find out and proceeded to pick up the phone on his desk and hand it to Jim. Jim called Phil and discovered that it had already happened. Phil had just finished speaking with Bob, and Bob had accepted the job.

"All right," said Harold matter of factly. "Let's wrap this up and get on with business. First, stop the transfer process. In all probability, you will have to explain to Bob why he must stay in his current role. You may have to sweeten the pot for him, given that you've already offered him a higher paying position. Secondly, I'd like to have you and Phil's plan for giving Bob greater satisfaction and insure he's a relatively happy trooper. When you have all that settled, I need to know your plan for Aklin and any other changes in assignments you're planning."

Jim confirmed he would do what Harold had asked before turning to leave.

"And, by the way," remarked Harold. Jim stopped, turned slowly and faced Harold, waiting. "Don't let something like this happen again."

Back in his office, Jim felt like beating his head against a wall. Self-doubt and suspicion about his ability as plant manager overshadowed the elation he had felt yesterday. He had disappointed his boss and, in the process, put himself in the embarrassing position of having to derail one of his first important decisions. He knew he was going to look like a fool with Phil and others. He wished he could disappear and avoid what was coming.

He was back at ground zero. He was less than two days away from a scheduled meeting with the entire workforce, and he wasn't prepared. He had stepped out on a limb and promised Aklin he would come up with a plan that would fully satisfy their needs and salvage their business. Now, it was back to the drawing board regarding just how he was going to accomplish that. A thought came to him. He went to his office and called Frank Zimmer, the plant manager for whom he worked just before taking on his new role. Frank had served as Jim's mentor over the years.

"Frank, I need your advice. Since you're about 50 miles away, could I come over this afternoon and have dinner with you? I'll buy," he said cheerfully.

"Sure," replied Frank. "Be good to see you again. Fortunately, I don't have anything on my calendar this evening. I need to check with JoBeth, but that shouldn't be a problem."

Later during dinner, Frank listened patiently as Jim told him how he had gone about starting his job and about the general problems he was facing with quality and delivery in the commercial sector. He covered the key position strengths and weaknesses he had analyzed and then spoke about Bob Frisman with whom he had apparently made a great error. As his story wound down, he reminded Frank that he respected his judgment and was hoping he could perhaps provide him with advice as to how to proceed.

"Jim, I'm honored you feel I can help you, and I'll do the best I can. However, you must promise me you'll take what I have to say as positive reinforcement and not criticism."

"All right."

"Good. Let me start with a question. What would you and your boss do if Bob were to suddenly drop dead?"

"To keep the business running, we'd do whatever it took."

"Correct," replied Frank. "You and your boss are focusing on Bob too much. You make it sound as if he's the only person who is capable of doing something constructive, and by your own admission, Bob isn't happy. So, you're probably going to lose him anyway—at some point. More important, you're overlooking a key player in all this."

"Really? Who might that be?" asked Jim.

"Joe. What's his name? Your production manager."

"Joe Thompson?"

"That's right. Joe Thompson. You see, if I were him I would be feeling just a bit left out about now. And, if he isn't, he probably isn't worth his salt."

"How's that?" replied Jim.

"Well, he is your production manager and I haven't heard you mention him playing a responsible role in all this. If you're having problems with quality and delivery, you should be all over his case about it. If he has people working for him who are causing problems with customers, then get him to resolve it. For instance, have you considered Joe taking over full daily production responsibilities for Aklin and relieving him of other duties until he has it straightened out?"

"No, I haven't," Jim admitted, but he found the suggestion intriguing.

"Have you taken time to pull your staff into an off-site working session, to address the entire Aklin situation and your business and to develop a collective action plan?"

"Again, no, at least not yet."

"Since they are a constant thorn in your side, have you taken the time to meet with Chuck and Cindy personally, to feel them out and get their thoughts about the problems?"

"I think I'm beginning to get the point," stated Jim.

"If what happened with your boss is the most horrible experience you ever have as a plant manager, then consider yourself lucky. It's not the end of the world, and you'll be faced with bigger obstacles and challenges before it's all over. Your boss is not all that upset. However, he's going to expect you to move forward in a positive manner. So, just learn from the experience, keep your head up and press onward. You've got too much talent not to end up being a successful plant manager. I applaud you for seeking someone out who's been there before. It just serves to reinforce what I said about you."

Frank went on about his beliefs and philosophy regarding plant management. He mentioned that in listening to Jim's story, his plant needed improved

manufacturing practices. He encouraged Jim to focus on that and keep himself in a position where he didn't get wrapped up in all the details. Though Jim
must establish an effective means of understanding the daily problems, he
should not be the active arbitrator in every matter. Jim's principal job, as plant
manager, was as a forward thinker. Stay above the small details, whenever
possible, and focus attention on the big picture, Frank encouraged. He finished stating Jim's chief responsibility was to establish a direction, which
would insure the operation became competitive.

"You won't be able to do this if you get sucked into the details and problems that surface on a daily basis. That's why you have a staff of responsible
and qualified people to handle those issues," Frank suggested.

Jim felt good about being smart enough to call Frank and learn from him.
Frank had this way of taking the most difficult problem and giving it a positive slant. What Jim was coming to understand, however, was that it really
wasn't a special talent that only Frank possessed, as Jim had imagined in the
past. Rather, Frank consciously rose above his self-doubts and frustrations
and looked at things with a fresh perspective. Jim intended to practice that
in the future.

The next day he asked for some time with Harold. Now sitting opposite
him at the small conference table in Harold's office, he got right to business.

"First of all, Harold, I want to apologize for what happened. You were
absolutely right. I made a serious error in judgment, but I believe I've learned
from it. I'm here because you asked me to get back to you on what I intended
to do to straighten things out. Well, I thought about it last night and here's
what I'd like to offer."

Jim explained he would like to resolve the matter without making himself
or anyone else look like the bad guy. Harold fully agreed and encouraged such
an approach, if possible. Jim then outlined his plan of action.

First, he would meet with Phil and tell him he and Harold had discussed
the matter and had decided to keep Bob in P&S. Instead, they would give
him a unit manager position. Jim wanted to make Bob unit manager over
P&S, reporting to Joe Thompson. Joe had seven individual foremen reporting
directly to him, and this was spreading him far too thin. Jim wanted to
unload some of Joe's current duties to give him more time to focus on
the Aklin account and the commercial sector and Bob would be ideal for the
role. Jim could tell by Harold's expression that he liked what he was hearing.
Jim continued.

"Joe is production manager, and it should be his responsibility to bring
things back in line on the Aklin account. In fact, it would be doing him an
injustice, otherwise. I'll, of course, work closely with him in formulating a plan
of action, but I want him to carry the ball. I'm going to suggest Joe spend most

of his time on the Aklin line, working with Chuck and Cindy. Meanwhile, I'll get my staff together for a general working session on our plant-wide problems and get their input for what I will say to the workforce. Because of these delays, Phil and I will set the workforce meetings up for early next week."

When Jim paused for a moment, Harold remarked, "I'm impressed, Jim. I have to admit I was little worried after you left yesterday, but it's apparent you've done your homework. I like your plan, and you're right about Joe. As production manager, he needs to be much more involved in resolving the problems with Aklin."

"It's a win-win situation. Bob gets the promotion, which he deserves, and Joe gets some relief so he can improve the commercial side and the Aklin account."

"I agree. Let's do it," remarked Harold.

The following Tuesday, Jim met the first group of employees. The meetings were scheduled to run on hourly intervals throughout the day and into the evening. His staff had worked with him over the weekend in preparation. Since it was Jim's first communication session with the workforce, Phil did the honor of introducing him. He stood briefly before the podium before looking up and scanning the audience of a hundred hourly associates.

"I'd like to say it's a privilege for me to have joined the team, here at Denning Industries. I'm honored to have been given this opportunity and I'm looking forward to working with you in the future. Though I'm new to the operation and we're getting to know each other, I'll bet I can guess what you want and expect from me. But, let me test that."

Looking over the crowd, he could see some appeared slightly puzzled by the statement.

"You want a plant manager who cares about you and about what's best for the operation." He saw someone nod in agreement. "You want someone who is willing to put in the time and energy required to make that happen." More nods.

"You want a person who treats people fairly." Even more nods. "You want someone to provide the direction required to take this operation forward to the front of our industry." A great many more nods and even an "Amen" from someone in the audience.

"Well, that's quite a challenge, but I can assure you I will, with all my might, do just that. That's my commitment to you, and I want to assure you that my staff, sitting behind me, is ready and capable to make this happen." Some slight applause.

"As you all know, Denning Industries has had a long and successful record. However, we are facing some serious challenges. Over the past two years, we've lost three major accounts. In every case, the customer expressed much displeasure in our delivery ability and product quality. But, I'm a firm believer that if an operation is in trouble, that trouble starts with management." The audience interrupted him with resounding applause.

"On the other hand, I would be remiss if I didn't point out that when the customer complains about poor quality, it generally comes from poor workmanship. That, my friends, is a direct reflection on you." The smiles slowly faded and heads stopped nodding.

"I think we all want to be the best in our industry, and I believe we can."

Silence.

"However, we face difficult challenges in achieving that reputation. Believe me, many of our customers feel we're far from being the best but, again, I believe we can make it."

More silence.

"It's going to take teamwork. As management, we have to listen better. We have to work harder to get your thoughts and ideas and do everything within our power to utilize your considerable knowledge and abilities. I believe we can." Someone clapped.

"In turn, you have to be willing to give management a chance to make a difference. That difference is going to mean changing how we currently go about our business, and I believe we can." Heads started to nod, again.

"Ladies and gentlemen, I've told you what I believe we can do. The key question is what you believe and what you're willing to support. A company is only as good as its employees. As management, we can lead, but success comes when you adopt a given philosophy and are willing to make it a success. It's more than a willingness to follow. It's a desire to participate in what sets us apart from everyone else. What will set us apart will either be something we collectively cherish and adopt as our motto or something that happens as a result of our inaction. We'll build a reputation, whether it's one we carefully plan and see through or whether it's one that, again, just happens. I, for one, prefer to have something to say about what the future holds and I hope you feel that way, too." Clearly, he had their utmost attention. He continued.

"I'm confident we can be the best. I'm confident we can take on any challenger and win, but I'm asking for your support and your participation to help us establish the proper course and make it a reality. That course can and should eliminate poor quality and meet customer demands, every time, without failure. What I'm suggesting is we have to change. Significant change, to say the least. Let me give you an example of the magnitude of change I'm speaking about. Have any of you ever been asked to work with inferior material or components?" More than one head nodded in agreement.

"Well that's got to stop and we're going to empower you now to refuse to do that." Some looks of surprise.

"Have any of you ever been asked to work with equipment that isn't functioning properly?" Again, heads nodded and a few raised their hands as a yes to the question.

"That, too, has got to stop, so starting now you're empowered to refuse to do that." Someone wanted to ask a question, but Jim requested he hold his thought until the question and answer period. He then continued.

"Have any of you ever been asked to build parts or components, knowing they were going to be reworked before they could be used?" A number of hands rose.

"That also has to stop, so starting now you're empowered to refuse to do that. In fact, ladies and gentlemen, we've prepared an official document, on a single sheet of paper, that I'm going to asked be handed out to you at this time."

He waited as preassigned individuals stepped forward and distributed the document. The auditorium hummed with conversation as more received their copy and began to read it. Then, he asked for everyone's attention.

"What you have before you is our new operating commitment. As you see, five items are listed, which I will allow you to read and think about later. Starting tomorrow, department managers are going to hold special sessions with their employees to cover this new operating commitment and answer any questions. Time's running short, and I have news I want to give you before we conclude this meeting. To begin, I am pleased to announce that Bob Frisman is taking a new position we've established, as unit manager over P&S."

Some immediate chatter went through the audience, along with a shriek of joy. Shopfloor employees liked Bob as well as did his peers, which Jim was pleased to know.

Jim politely signaled for everyone's attention. He let them know Bob would be reporting to Joe Thompson. Joe would spend much of his time on the Aklin line, in an effort to help get the department back on schedule and, more important, increasing quality.

Then, he broke the news Aklin was considering moving to a competitor, which apparently wasn't such big news by the look on everyone's face. Rumors travel fast in the manufacturing world, but they did act surprised when he told them about his invitation to visit with Aklin management and assured them he was going to do everything possible to salvage the business. He ended promising he would get back to them on the results.

Jim requested only business-oriented questions and that this was not the proper forum to discuss personal issues, such as grievances. The individual who had earlier signaled he had a question was the first to speak up.

"You say you're empowering us to refuse to work with bad material and with equipment that isn't running right?" he interjected, waiting for confirmation. Jim nodded.

"So, who's gonna pay for the groceries and our bills when we refuse and get fired?"

A thunder of laughter rolled through the audience and some turned to each other and began carrying on conversations. Jim raised a hand and patiently waited for things to quiet down.

"Your name, please?" Jim requested.

"Adams," he replied, "Jock Adams."

"Mr. Adams, that's a fair question, but we are serious and we will live up to our end of the bargain. You see, change is a two-way street. We're going to ask you to change the way you've done things. In some cases, this will be frustrating and will cause confusion as we sort through it all. In turn, as management, we must demonstrate we're also willing to change, which explains the document you have before you. It's our solid commitment. As to your question, I'll say this. Every person standing in this room has my personal assurance that if he or she abides by the document's guidelines, there will be no retribution against anyone, at any time. You have my word on that."

By the time the day ended, Jim was exhausted. He had just finished conducting his eighth and final meeting, with third-shift employees, when he heard a knock on his office door.

"Yes?" Jim said.

A well-dressed woman, in her early to mid-thirties, stepped into Jim's office.

"Excuse me," she said politely. "My name is Cindy Mabrey. I was hoping you might have a few minutes?"

"Oh, yes. Cindy. Well, nice to meet you," he replied, stepping forward and shaking her hand. Won't you have a seat?" Cindy sat down and folded her hands in her lap, waiting as he walked around his desk and took a seat facing her.

He smiled, "It's been a long day for me."

"Oh, I understand. Really. I was just hoping you had some time. It doesn't have to be now, you know. It's getting late. I mean, when you have some time."

"No problem," he replied, sitting back in his chair and smiling, for added assurance that he was ready to take the time now. "What's on your mind?"

"First of all, Chuck or Joe have no idea I'm here, and I would prefer to keep it that way."

He decided not to encourage or discourage her, so he waited for her to continue.

"The news is out about the Aklin account with Joe coming to take over. Chuck says Joe's been told to straighten it out pronto and if not that he and I will be on the street."

"Oh?" replied Jim.

"That's the talk going around, and I'm here because it has me worried. I'm a divorced mother with two children and this job means everything to me."

Jim was too tired to be as diplomatic as he might have been under different circumstances, so he got straight to the point.

"Cindy, there's been considerable discussion about the area's problems and negative feedback I've gotten regarding your interface with others, along with Chuck's. That's why I asked Joe to get involved. Not to build a case against you, but to provide you with solid advice and direction. You could be making a mountain out of a molehill. I say that because everyone is on equal footing at this point. What's important to me is not your past record, but your future performance. If you and Chuck or anyone else prove your worth, then you have nothing to worry about." Cindy sat quiet for some time, never taking her eyes off him. He didn't know what to expect, but he waited.

"Mr. Warring," she said, "that's all anyone can ask. Thanks for being honest. Most in your position wouldn't have done that. But I can assure you I'm going to go back and do the best job I can."

"Well, I'm glad to hear that and I'm sure you'll do just fine," he remarked.

After she left, Jim sat for a few minutes, thinking about the day's events. This job was going to be more of a challenge than he had expected. One last time, he scanned the document that his staff had distributed to everyone in the communication meetings. He was proud of the work he and his staff had done. Across the top in bold letters, it read:

DENNING MANAGEMENT COMMITMENT

After a brief statement came the list of five specific items.

Denning Management is committed to providing a world-class manufacturing environment for its employees. Therefore, the employees of Denning Manufacturing are empowered and fully encouraged to refuse to perform work, when one or more of the following conditions exist:

1 Any assignment that requires the use of defective raw materials.
2. Any assignment that requires the use of defective purchased components.
3. Any assignment that requires the use of defective parts from other plant areas.
4. Any assignment that requires the use of defective equipment and/or processing.
5. Any assignment that is clearly generating scrap and/or rework.

WARRING SCORECARD, PART 1

So far, Jim Warring has done a number of good things, but the measure of an effective plant manager isn't as simple as looking at positives verses negatives and establishing a score. In fact, most managers would perform decently based on that criteria. The true measure is the number and depth of crucial mistakes in judgment.

A few relatively serious mistakes over the average plant manager's tenure can be acceptable, as long as these don't point toward a lack of good judgment. If this happens, the manager could lose his or her job and a potential future in management, which is why the position can be frustrating and why good plant managers usually aren't timid.

Throughout this continuing scorecard, we will examine how Jim Warring is doing in two important job aspects: As a plant manager and how his actions as the "chief conductor" for lean engineering served to either advance or restrict the plant's overall progress.

As Plant Manager

1. He studied key positions and obtained valuable input from other functions.
2. Even under pressure, he delayed his first meeting with the work-force until he had the facts, knew about the operation's competitiveness, and involved his staff in outlining a direction and commitment.
3. When he was in serious trouble and needed help, he sought out a qualified mentor, listened carefully, and followed his advice.
4. In dealing with a surprise meeting, with a high-ranking member of Aklin's management, he reacted in a responsible and effective manner. Therefore, he had the opportunity to take one additional step to salvage the relationship.
5. When he was wrong, he admitted it and established a win-win plan of action.

Jim made one serious judgment error regarding his freedom to make change. He moved Bob Frisman after Harold had informed him he would not support it. Harold's conversation in the car led Jim to believe he had unlimited decision making freedom.

However, other less obvious, yet potentially serious, errors include the following:

1. Not listening closely enough to at least one of his key players.
2. Promoting an individual to a leadership role without first consulting another manager.
3. Charging an individual with cleaning an area up without knowing his ability.
4. Being unclear with written documentation.
5. Making crucial decisions affecting the workforce without involving the union.
6. Not appropriately recognizing and communicating with the extended staff.

In the upcoming Warring saga, these will serve to haunt Jim. As you will see, being a good plant manager doesn't mean having perfect judgment or making the perfect decision. It depends on how well one responds to the unexpected and the unknown.

As Chief Conductor for Lean Engineering

Jim's greatest achievement was the creation of the Denning Management Commitment, which empowered employees not to perform work under certain operating conditions. However, Jim allowed a golden opportunity to slip away by not driving home the absolute need to adopt and pursue the implementation of lean manufacturing practices.

In his initial tour through the plant he saw examples of a conventional manufacturing operation with excessive wastes. His first meeting with the workforce was the ideal time to have shared his observations and his convictions to improve their competitive edge.

3

Setting the Course

SHORTLY AFTER ARRIVAL, the workforce will be looking for the new plant manager to set a course or direction. A plant manager should always approach the job having a specific focus in mind. Otherwise, getting the job done in an effective manner will be more difficult. Focus can and will vary, but it is largely dependent on the degree of competitive health, customer satisfaction, and employee satisfaction.

Competitive health includes the operation's market share, profits, etc. Customer satisfaction is how customers perceive a company's ability and flexibility to deliver its product. The higher the flexibility, the greater the customer satisfaction. Employee satisfaction centers on job security and employees' perception of working conditions and the relationship between company and union, should one exist. See Figure 3.1.

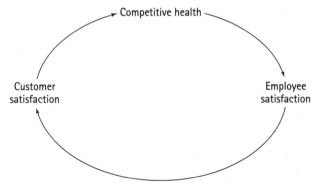

Figure 3-1. Three Factors of Operational Success and their Dependent Relationship

If you find an operation with all of these in good working order, the focus should be on forward planning. This means those things that will help the company expand business and enhance profit levels, e.g., focus on new and advanced manufacturing technology.

However, having all of these in good order is normally the exception. When inadequacies exist in one or more of the above, they tend to exacerbate each other. Therefore, poor competitive health can lead to poor customer satisfaction. In turn, poor customer satisfaction can lead to poor competitive health and dissatisfied employees. Dissatisfied employees can have a negative impact on customer satisfaction and the operation's competitive health.

I did not include the *financial health* of an operation as a key factor because it is proportional to the three items noted above, which tend to feed off one another like a slow cancer. Sometimes this erosion is so slow that when a company decides it has a serious financial problem, its customers may have already irreversibly decided to take business elsewhere, while the root problem may have been worker dissatisfaction.

It can and does happen, and you would be surprised how often. You might also be surprised how frequently at least one of the three is taken lightly by plant managers. A plant manager once told me, "I'm not concerned about employee satisfaction. I pay my employees better than ten dollars an hour to be satisfied and if that doesn't do the trick they're more than welcome to search elsewhere for it." He no longer holds that particular management position, but he was viewed as highly successful until his operation almost went under, due to new competition coming on the scene that behaved quite differently.

In my younger days as a manager, I too felt what pay and benefits a company provided its employees should be enough to earn their enduring loyalty. I had to learn the importance of never underestimating the competitive advantage of highly satisfied employees. Conversely, I discovered the destructive power of a dissatisfied workforce. I corrected my ways before it ruined my career, but it did not come naturally. Like venturing into deep water for the first time, once you realize you're over your head, you instinctively learn to swim or you drown unless someone is there to rescue you. But in the vast sea of business, lifeguards aren't usually standing around waiting to help.

New plant managers must start things properly, otherwise they end up reacting to the arising circumstances. Appropriate reaction to the unexpected and unknown is important because it is the position's reality. Underneath all that should be a driving motive for existence. In other words, why are you there and what do you want to accomplish? Keep this in mind at all times or you could send a confusing message to the workforce.

Assume a plant manager has been hired and dissatisfied customers are apparently the number one problem to overcome. Then, consider the solution calling for extensive overtime to meet customer demands. At the same time, upper management is applying considerable pressure to reduce operating expenses, in order to meet financial forecasts. What does a plant manager do? The answer is you schedule the overtime and take your lumps regarding expenditures.

Some plant managers, under the same circumstances, would let their people know they had to check with the powers above before making a final decision. Even if their answer is go ahead and work the overtime, this sends the wrong message about where your priorities lie. It strongly implies you are afraid of making your own decisions or that you are truly not in command. Either way, the message will do little in creating the proper respect for you.

Those with manufacturing experience have heard some say they could easily do the plant manager's job because they have seen too many instances where plant managers indicated they really were not in charge. These managers have left the impression their principal task is to carry issues and problems forward to the powers above and then guide the implementation of such decisions. I call this the messenger boy syndrome.

Still, plant managers should not charge ahead and completely ignore the direction and counsel of those above. They should analyze all reactionary decisions they are asked to carry out, with respect to the prime objective they want the workforce to implement and the impact those decisions could have.

Effective plant managers must express strong convictions about how they intend to lead and where their priorities rest. Without this, they will be perceived as merely messenger boys, and their ability to get the workforce behind their vision will be severely limited.

THE WARRING ADVENTURE, PART 2

Jim listened as Joe Thompson informed him the workforce had given some positive feedback about the communication meetings, specifically regarding the management document and Jim's comments about it.

"Some are still a little skeptical," he pointed out.

"I'm sure," replied Jim. He waited, then proceeded to change the subject. "How are things going with your new assignment?"

"Oh, fine," replied Joe.

"I'm looking forward to seeing your plan of action, Joe. I'm going to need this and something from Bob Hampton quickly, in case I do get to visit Aklin." Bob was the operation's quality assurance (QA) manager.

"Of course," he replied, "I'm working on it."

"Any chance of reviewing this tomorrow?"

"Tomorrow?" replied Joe, surprised at Jim's proposed time frame.

"Yes, if possible. I don't expect it to be finalized, but I need your thoughts on it."

The phone rang. It was Fred Johnson. "Jim," said Fred, "the meeting is set up for next week, the seventeenth. I told them you would be there, considering its importance."

"Of course, no problem," Jim responded.

"The entire management team at the Camden facility will be there. When you have time, I'll tell you about the key players and what you can expect. I told them you might bring Harold and a few of your own people. Have you considered taking Bob Hampton?"

"I haven't decided yet. I'll get Linda, my secretary, to set up a good time with you." With the conversation over, Jim turned his attention back to Joe. "Speak of the devil. Fred tells me the Aklin meeting is set for next Tuesday, so we have to hustle. I want you and Bob to discuss this and then get back to me about three this afternoon so we can start laying out a plan."

"Fine," replied Joe and then left.

As a result of Jim and his staff drafting the Denning Management Commitment document, he was proud knowing it was a joint effort of the entire management team. However, he knew Aklin would need more than a promise. He had to have something new, different and possibly unique, if Aklin would seriously consider Denning keeping their business.

Harold had made it clear: Though he supported Jim visiting Aklin, he didn't give him much chance of changing Aklin's decision. "After all, you're new here and the end of their contract is only six months away. You probably won't be able to make a difference in that time." Harold had given up on the Aklin account and was placing his focus on other business matters, but he had strongly encouraged Jim to put his best foot forward because "it could make a difference in getting another Aklin account in the future."

Jim wasn't so sure Harold was right on this one. If they had already made a hard and fast commitment to another supplier, then Jim wondered why Cummings had encouraged his visit. He needed to know more, so he called Fred Johnson and went right to the point.

"Fred, I was wondering how you'd feel about me calling the Camden plant manager, introducing myself, and feeling him out about the upcoming meeting?"

"You mean Jason Andrew. I don't see why not," he replied. "But let me warn you. He's one tough cookie. I mean, tough. I've seen him in action. If you do get the chance to speak with him, be prepared for him to unload about all the problems he's had to endure. In fact, I was going to mention that to you when we got together later."

"I can handle it," replied Jim. "I was wondering: Who calls the shots on their suppliers? Jason or someone else?"

"Someone else definitely," responded Fred. "Their corporate purchasing group, in Atlanta, does that. Carol Hamilton heads that end. Her group does all supplier negotiations."

"I see," replied Jim. "If that's the case, why the meeting with the plant group instead of Ms. Hamilton or someone in her group?"

"Believe me," responded Fred, "she'll have someone at the meeting."

"So, you don't think it's too late to salvage the business?" asked Jim.

"Oh, without a question, it's too late."

Jim paused, not understanding what was going on, and then decided to be more direct. "Then I don't understand why I should waste time going to a meeting to take a lot of unnecessary heat if I have no chance of keeping their business."

"Didn't Harold tell you?" Fred asked, sounding surprised. "Aklin's been awarded a multi-million-dollar contract and will be subcontracting the panels to one or more suppliers. It'll be a major account, and we'd like to have our foot in the door on it. In fact, we've already discussed specifications with them, but they've given us a bit of the cold shoulder. They like our cost and service—we're competitive there—but they're afraid manufacturing won't deliver. Of course, the workmanship stinks in their judgment. So, we were hoping you could convince them to give you a chance to make a difference."

Aklin made wiring harnesses and often sealed contracts for the decorative panels, which covered the wiring, because they did not produce such panels. As a result, Aklin often served as a first-tier supplier to the aircraft makers, and Denning produced components Aklin and others used in completing their assemblies.

"Well," said Jim, "thanks for clearing that up."

Fred agreed Jim should call Jason Andrews, which he felt would be a good idea. Jim thought through their discussion. He was disappointed Harold had failed to be as clear about it as Fred. He wondered if Harold was up to speed on everything though Fred was surprised Harold hadn't already told him the news. Jim was confused, to say the least.

He almost decided to call Harold to clarify the matter. Instead, he put the phone back and decided to wait. He needed to prepare for his meeting with Joe Thompson and Bob Hampton.

Later, sitting with Joe and Bob, he began. "Gentlemen, what we need to do today is decide what plan and commitment I am going to offer Aklin," said Jim.

"Outside of offering perfect quality on every unit we send them, which we probably can't do, then I'm at a loss on this one," replied Bob.

"And you, Joe?" asked Jim.

"I feel much the same way," he replied.

"Let's see if we can take a slightly different approach," said Joe. "Bob, isn't poor workmanship the biggest quality problem with Aklin?"

"Yes," responded Bob. "I'd say at least 95 percent of the problem."

"If that's the case, what could we do to prevent that from happening, and don't limit your thinking on this."

Bob paused for a moment before responding. "Until we find a way to build quality at the source, the only way to reduce defects would be to perform a 100 percent dock audit. This means inspect all goods after they have been packaged and boxed for shipment."

"And that would be ridiculous," interjected Joe.

"Have you ever conducted dock audits in the past?" Jim asked.

"I have, but it was with another company who commonly performed dock audits if the customer had problems. I've never heard of a dock audit on a continuing basis. That would be an expensive proposition, to say the least. A 100 percent inspection still doesn't yield 100 percent error free product."

"I understand that," replied Jim, "but how about 200 percent?"

"You mean re-inspect goods twice prior to shipment?"

"That's correct."

"Though it would be twice as good you still couldn't guarantee perfect quality."

"What I was thinking," said Jim, "was a procedure with a 100 percent dock audit followed by a second complete inspection if rejects reached a certain percentage. We would include some water mark where the first inspection would be cut back to a lower percentage if we statistically demonstrated it wasn't necessary."

"That's an interesting concept," Bob admitted.

"Excuse me," Joe interjected. "Are you saying we're going to consider doing something like that?"

"Perhaps," replied Jim. "We have to free our thinking here, Joe. Think of the alternatives. We won't impress Aklin with promises to improve unless we can show definite actions."

"Well, we've lost the account anyway," he started to elaborate.

Jim interrupted Joe and related what Fred had told him. "What we have is possible future business with them, and now is the perfect time to demonstrate we mean what we say, up to and including an intensive dock audit, if required."

"I understand," replied Joe, "but you can't run a business like that on a continuing basis."

"Joe, I'm not suggesting that we put such a procedure in place and then forget about it. As Bob mentioned, we have to build quality at the source, which we will address. In the meantime, we have to show Aklin we will do whatever's necessary to provide them with good quality. I don't want just good quality. I want outstanding quality." The three men contemplated the possibilities. Jim continued. "As I see it, the idea of a dock audit, coupled with actions to make this operation a world-class manufacturing facility, should get Aklin's attention. Then, we have to do what we say."

"But, how about the cost of all this?" asked Joe.

"Let me worry about that, Joe," Jim replied sternly. Joe lifted his hands in a gesture of frustration or disgruntlement. Jim wasn't sure which, but he knew he had to learn more about Joe's inner convictions and priorities. He decided now wasn't appropriate.

HOW TO DEVELOP AND EFFECTIVELY EARMARK YOUR PLAN

Your plan must go beyond some general philosophy. I have seen beautifully worded plaques on the walls of factories theoretically announcing how the company intended to conduct business. However, when push comes to shove, these plaques seldom represent how the facility operates. To be effective, your plan has to be based on the operation's specific conditions and needs, and it must have a strong demonstration of conviction.

Some might see this as approaching a race with blinders on and, to some degree, that's correct. On the other hand, you can be flexible to changing circumstances while keeping strong convictions. It means no matter what's going on around plant managers, they have to stay focused on the primary mission they have adopted and communicated. For a process to develop and earmark your plan, I have a few points I would like to offer:

- Begin by fully understanding the specific needs of the operation. Do not let past convictions mar your creative ability. For example, if you use lean engineering concepts and techniques, ensure you have the proper foundation or confusion will occur and you will be unable to implement the process quickly and effectively. When a ship is sinking, the captain should not form a taskforce to measure the water gushing in. It's time to seal the hole! However, such emergencies do call for some excellent knowledge about what is making the operation less than successful. This must be done without preconceived notions or perceptions getting in the way; therefore, obtaining the facts, and only the facts, is essential.
- Consider short-, intermediate-, and long-range initiatives. Your short-range plan should focus on resolving immediate customer satisfaction issues. If not a problem, then proceed to the intermediate level and emphasize employee satisfaction and/or the application of improved manufacturing techniques, e.g., waste-free manufacturing. The key is to build a foundation allowing continuous improvements, with employees who like what's going on. Finally, look at long-range initiatives that will put the operation in good competitive health. This specific focus would definitely include the implementation and practice of waste-free manufacturing or lean manufacturing techniques.
- Back up your plan with constant communications, training, and emphasis. Get the message out and keep it out! Do not communicate this once or twice to the workforce and hang some nice posters or plaques throughout the factory. It has to be lived and practiced daily by the plant manager and his staff. Every written or spoken communication must mention this, particularly those from the plant manager. Every event needing a reaction by the plant manager and his staff must highlight its potential impact on the mission.
- Carefully select those helping you champion the process. Every plant manager needs dedicated people, in key positions, who make the plan their personal vendetta, preferably a direct report and a

staff member. If you do not have them, find them quickly. If you must remove someone less loyal to the mission, then do it. Distractions will swamp you everyday. So, you need people who can dedicate the needed time and energy to the welfare of the plan or it will probably not happen.

- Then just do it. After setting the above four actions in place, implement the plan. Avoid revisiting the plan repeatedly to comfort yourself. Any worthwhile plan will provide some discomfort, along with certain risks and uncertainties.
- Most important, it is a plan, a blueprint, and can have corrections and revisions as long as the mission's core remains intact. Once developed, people may feel a plan is cast in concrete and the unpardonable sin becomes revising it. They tend to become blind to the real mission, which is meeting the plan's basic intent. Instead, they see change indicating shortcomings and measure success on how many of the original steps were completed. For instance, consider our country's mission to put a man on the moon and return him safely to earth. Circumstances could have occurred preventing Neal Armstrong taking that historic step. If he had been unable to do so, mission control would have immediately switched the plan and used his co-pilot, but the core mission objective would not have changed.

Jim had asked Joe to stay after their meeting with Bob. When Bob later departed, Jim got right to the point.

"Joe, do we have a problem?"

"Problem?" he asked.

"Yes. Evidently, something I said or how I said it bothered you. I would like to know where we stand on that. I don't expect you or anyone else on the staff to agree with me on everything. I'm not looking for blind compliance, but I would like to know of any serious conflict with my position on this matter. That's why I have you here."

"Well. Yes, I do disagree on a couple of issues. Perhaps I should have said I have a concern rather than any strong disagreement."

"That's fair enough," replied Jim. "Could you share it with me?"

"You and Bob spoke about inserting costly audits. I interjected my feelings but got cut off. I'm concerned about where we get the people for that. Number two, how am I to meet my stated financial objectives if we pursue such an approach? I won't beat around the bush. My annual bonus is based on how well I perform against those objectives."

"I see," replied Jim, pausing to consider Joe's comments. "Well, Joe, we have more people than we need. I've spent enough time on the shop floor, enough to know what's going on. On several occasions, I found fewer than 20 percent of the people performing value-added work. Regarding your objectives and the impact on meeting them, are you telling me MBO's are cast in concrete and never change, even when conditions and assignments change?"

"That's correct," he replied.

"They're never changed or revised?"

"Seldom."

"Well, we'll have to correct that," said Jim.

"Good luck," replied Joe.

"Don't keep me in suspense. Why not?"

"It's a long story, but the MBO procedure comes out of HR and they don't change established goals and objectives once they're set."

"That doesn't make sense. How then do people who change assignments get measured?"

"They usually don't," Joe replied in somewhat a sarcastic manner.

"I'm confused at this point. I'll look into that and get it corrected so it should have no bearing on your or anyone else's wallet. Don't worry about it, but regarding the plan we discussed with Bob, how do you feel about my comment on the number of people used?"

"Oh, I mostly agree we have more people than should normally be required."

"I'd like to get back to that when we have more time, but thanks for hearing me out and for your comments," Jim said, extending a hand to Joe.

Joe looked confused that Jim was abruptly ending the conversation, but he stood and shook Jim's hand firmly before leaving.

"By the way," Jim said, causing Joe to stop and turn around, "has Carol said anything lately I should be aware of?"

"No, why do you ask?"

"I was just wondering. Thanks again."

On the way home, Jim ran the conversation with Joe over in his mind. He was expecting more of an unselfish commitment to what was best for the business. On the other hand, Jim had been around enough to know many things affect objectivity. Jim told himself he shouldn't jump to conclusions at this point.

Something about Joe was troubling him. Here was Jim's top manufacturing manager (in terms of overall authority and responsibility and the number of direct reports) who was the least active in his staff for the off-site planning session. Bob and Phil had outweighed Joe for ideas and contributions. Zack

Milen, the engineering resources manager had also impressed Jim. In fact, Jim felt reasonably good about the rest of his staff.

Again, something in Joe's general demeanor and his response made Jim uncomfortable. He had to settle it, one way or the other. Joe held such an important position that Jim simply couldn't hold doubts about his abilities. If it was simply a personality trait, Jim could live with that as long as Joe got the job done. But if it were anything else, he would have to make a change and do it quickly.

The following day, Jim had just finished touring the factory and returned to his office when Linda told him Carolyn Hamilton called and asked him to get back to her as soon as possible. Carolyn Hamilton was Aklin's corporate purchasing manager Fred Johnson had mentioned. Jim returned the call, introduced himself, and asked how he could help.

"Mr. Warring, I've been made aware that you're going to visit our Camden facility next week, the seventeenth and will be putting on a presentation."

"That's the plan, as I understand it," replied Jim.

"Would you mind telling me about your presentation? As you know, we've negotiated a contract with another supplier and they will get our business at the end of the year."

"Yes, unfortunately I've heard. But I was planning to provide some input about our plans to improve our quality and delivery."

"That's all well and good, but you need to understand the matter is finalized. We simply can't back out of our contractual obligations."

"I fully understand that," replied Jim. "My focus will be to show Aklin we're on the right track and should be strongly considered for future business."

There was silence on the other end for what seemed an eternity to Jim.

"I don't know where you've gotten the idea Aklin would consider doing business with Denning again. That just isn't going to happen."

Now Jim was at a loss of words, but he quickly gathered himself.

"You're telling me there's no chance Aklin will ever do business with us again?"

"That's correct," she replied.

"I'm not being sarcastic, but how can you make such a statement? Never is a long time."

"Well, it's never as long as I'm in charge of corporate purchasing," she replied quickly.

"Obviously, we've seriously upset you and for that I apologize, but it was before my time. All I'm asking is the opportunity to tell you and others at Aklin what we intend to do to become the best in our industry. Once done, if you feel the same way, then so be it. We'll pull up our stakes and move on."

She was quiet again, obviously thinking about her response. "I want to be fair because you sound like a nice enough guy. We've had so many problems and unfulfilled promises with Denning that our patience has reached its limit."

"Again, I understand, but I need a chance to show we will go beyond the promise stage."

"Oh?" she replied, genuinely interested. "How will you get beyond the promise stage?"

"Give me the chance I mentioned and I'll tell you," he answered.

"All right," she finally agreed, "the meeting is already set, so it saves me the trouble of calling everyone to cancel it. I have to advise you, Mr. Warring, I'm going to be there to hear what you say. I hope you don't mind if I challenge you when a question or concern comes to mind."

"Not at all. It's good to know you'll be there, So, we'll see you on the seventeenth?"

"I'm looking forward to it," she replied.

Following his conversation with Hamilton, Jim called Howard but had to leave a message on his answering machine. Jim called Fred, who happened to be in his office.

"Guess what?" Jim said. "I just got a call from Carolyn Hamilton."

"What about?"

"She called to find out about the meeting in Camden. Can you believe it? I would have thought she would be the first to know. She came right out and said there was no chance of them giving us any future business. Period."

"She did?"

"Yes, but I convinced her to hear me out on the seventeenth, so it's still on."

"That's good," Fred replied. "My guess is Jason, the plant manager at Camden, hasn't told her. If you think we have problems, you should see their internal communications problems. Jason doesn't like anyone at headquarters and he's an independent cuss. When you speak with him, you'll understand. I'm glad you were able to convince her to come."

"She sounded as if she absolutely abhors us. Perhaps there was some advantage in me being new to the job. Anyway, we'll see how it goes."

No sooner had he hung up the phone when it rang again. It was Harold, returning his call and Jim told him what he had told Fred. Harold appreciated the way he handled things, overall. They agreed to discuss what Jim had in mind for his presentation. Jim sat back for the first time that day to relax for a few moments. He felt good about the way he had dealt with Hamilton and took some pride in knowing both Fred and Howard seemed pleased with his actions.

The phone rang again. Linda took the call, and she came to his door. "Mr. Warring, Bob Hampton's on the phone. Says it's urgent." Bob was Jim's QA manager.

"Jim, I'm calling because we have a level four quality alert on the Aklin line," Hampton advised.

"A what?"

"A level four alert is when we are forced to put all product for a given customer on hold, including anything they have at their plant. So, we have to call and let them know. In turn, we have an obligation to send people to their plant to inspect and rework our product, if possible. Then, we have 48 hours to get back to them on the root cause."

Jim tried to remain calm. "What's the basic problem?"

"The material we use on the panels we build for Aklin is cracking. We've experienced this problem, off and on, for the last few weeks but thought we had it under control."

"Wait a minute, Bob," Jim interrupted. "Are you telling me we've had a quality problem on Aklin products for weeks and you're just now making me aware of it?"

"I guess so," replied Bob. "While it's something you and I haven't specifically discussed, it's spelled out in the daily quality reports I forward to your office."

Jim took a deep breath and then sighed. It seemed that every time he made a little progress with Aklin, it was followed by an unexpected crisis.

"We'll get back to the quality reports later, but for now what's the bottom line here?"

"In two words, not good."

"I was thinking more in specific terms," said Jim.

"We could temporarily shut them down, unless we can inspect enough good product to keep going. This comes and goes; one roll of material will be good and another bad. We just never know. Performing a reliable receiving inspection is almost impossible."

"What sort of test do you perform on the material?" asked Jim.

"Outside of life testing, which is a time-consuming lab setup, the only thing we do is a pull test. You put a prescribed amount of tension on the material and if it tears, then it's probably going to crack later down the road."

"Is Aklin aware of the problem?"

"Oh, yeah. We've been working with them on it since they specified the new material."

"They specified a change in material?"

"Yes. We made the change about three months ago to a less expensive material."

"Now I'm beginning to get the picture. So, Aklin has to take some of the blame for this?"

"Well," he hesitated.

"Let me try this from a different angle. Have we ever had such problems prior to making the material change that Aklin specified?"

"No, none at all."

"Was the material tested, under all our standard testing procedures?"

"Yes and there were problems from the start," replied Bob.

"And?"

"Well, Aklin insisted we work them out. Their position was that we were the panel manufacturer and we should be able to find a way to resolve the problem."

"And we bought that?"

"Yes, I'm sorry to say."

Bob went on to explain he had sent a number of letters to Aklin engineering about the poor quality of the material, along with extensive life testing results. He noted he would send Jim copies of the report. Bob said Aklin had a history of making material changes after a contract was signed and delivered and, when problems occurred, the tendency was for them to place blame on their suppliers. When Jim asked who was driving the changes, he was surprised to hear him say it was Aklin's corporate purchasing group, which was headed by Carolyn Hamilton. Jim explained he had just had a conversation with her, but he would have thought the change was dictated by their design engineering group.

"Oh, no," he replied. "The Aklin engineering and manufacturing groups detest such changes. It causes all kinds of special problems for them. Carolyn and her purchasing group insist on the changes, which come as a result of a corporate mandate to lower overall material costs. I'm sure you've heard all this, already."

"Frankly, Bob, I haven't, but that's another story."

They agreed Bob would get all the key players together and meet Jim in his office. Jim saw one small ray of sunlight in all this. At least, workmanship was not the problem.

HOW AND WHEN TO SEEK THE UNION'S ADVICE AND SUPPORT

If the company has a labor union, there comes a time to seek the advice and support of those who hold leadership roles. Often, the plant manager should meet with the president of the union directly, but the meeting can include all the elected officers who serve as members of the Union Committee.

You shouldn't wait for a crisis to introduce yourself to the union management and start a close and continuing dialogue. Read *Fast Track to*

Waste Free Manufacturing, which explains how make the union a partner in change. In doing so, you should personally and candidly approach
them about ongoing manufacturing problems first. The idea is to solicit
their help and support on contractually uncontroversial matters.

Don't allow anyone to tell you dealing with the union is not your job.
Most often, you will hear this from those whose chief responsibility is
carrying out labor negotiations. They can be high-ranking officials at
headquarters, but they are normally at the local level and have dotted-
line responsibilities to one or more individuals who report to functions
outside manufacturing.

Regardless, you should work on building a personal working relationship with the union leadership. This gives you the chance to ask
and, under certain conditions, expect their advice and support. Such
conditions definitely include the following:

1. Change that has an impact on job duties and responsibilities
2. Change that has an impact on the quality of work life
3. Change that has an impact on seniority and job security

When one or more of these conditions arise, sit down with the
union leadership and get its input before proceeding with general communications. Though this will cover when to seek advice and support,
how to do it is more complicated. Here are some basic steps and rules
to keep in mind:

Step 1: Give the union news as early as possible. Insure the union
leadership is the first to know, regarding formal workforce notifications
on any matter. This can mean before some of your direct staff and
most, if not all, of your extended staff. The purpose is to build a relationship with the union showing your full respect for the job they were
elected to perform. Whether you admire, hate, or are indifferent to labor
unions, you are obligated to make the best of the relationship.

Step 2: Keep the union continually advised of all changes. One of
the more crippling occurrences is to work hard at step one and then
embarrass the union leadership by not advising it of a condition or circumstance that changes what it was initially told.

Step 3: Listen to what the union recommends. Under the job's tension and pressure, you can easily forget what was said. I have done this
and I always regretted it. To be certain, ask for absolute clarification

before the end of any meeting with the union. Though the members may have already mentioned it, they probably won't mind repeating themselves for clarity.

For anyone harboring negative feelings about labor unions, effectively dealing with them is like working with equipment you don't like and wish you didn't have to tolerate. Until the equipment is gone, you have no choice but to deal with it, which means proper care and maintenance. Personal feelings aside, if you display appropriate respect for the Union, you will have access to some of your operation's best-informed sources.

In addition to the above three steps, the following pertains to two important rules never to be compromised, under any circumstance:

1. Never surprise them. This goes further than steps one and two. Never surprising the union leadership means always being aware of their need to know. So, do to others as you would have them do to you. You should inform the union about most things involving employees. This can include something as innocent as a company party. In fact, we had scheduled a Christmas party for the entire workforce, and I inadvertently failed to cover this with the union leadership before the announcement was posted. I learned the date conflicted with some year-end festivities for international and local union members. As a result, the union was embarrassed when members asked them which party they should attend. Though we did change the date later, the damage was done. I learned from that incident and made certain it never happened again. Most important, it instilled a strong awareness that a plant manager should never take anything for granted when it comes to communications.

2. Never forget confidentiality. Plant managers must build a relationship so they can discuss matters openly with the union, without fear of something being taken out of context or that confidentiality will not be honored. In addition, plant managers must always be careful what they discuss with others, pertaining to union business. The most innocent remark can turn into a nasty and damaging rumor. So, the less you say to others about your union leadership conversations, the better.

Jim was worried. After his conversation with Bob, he learned the company had a strict policy that called for the plant manager to issue a Level Four

Quality Alert (ALQA). This demanded certain actions be accomplished, in precise order:

1. Let the office of the president know an ALQA has been issued and which customer(s) it will affect.
2. Advise the sales and marketing group, the product engineering group, and the corporate logistics group.
3. Contact the plant manager(s) at the affected customer site(s) and pass on any required information.
4. Within 24 hours, submit a copy of a Problem Identification and Information Form (PIIF) to everyone noted above.

Jim found himself thinking the only thing it left out was: "Cover it first with your boss, stupid."

After going over the basic details with Harold and promising to get back with him as soon as possible to discuss it further, Jim sat out to follow the policy instructions. He had to call Jason Andrew, not the type of call he had been planning to make. Plus, he couldn't help but remember what Fred Johnson had said about Jason.

When Linda informed Jim she had Mr. Andrew on the line, he took a deep breath and picked up the phone.

"Hello," said Jim.

"Yes," responded a deep, coarse voice on the other end. "Jason Andrew here."

"Mr. Andrew, I'm Jim Warring, the new plant manager at Denning."

"I've heard," he interrupted.

"I wish I could tell you I was calling to introduce myself and say hello, which was my intention, but instead I have some bad news I need to pass on."

"That seems to be par for the course for you guys."

"I'm sorry you feel that way, really, but I guess I should get to the point."

"Please do," Andrew demanded.

"I'm calling to let you know that we've been forced to issue an ALQA."

"Not again?" exclaimed Andrew.

"I'm afraid so. We've discovered the new material, which was recently specified by your design engineering group, is potentially going to crack in the field."

"Damn it! I tried to warn them about this."

"Who was that, Mr. Andrew?" Jim asked.

"Those dummies in purchasing. It's been a problem from the get-go."

"Yes, I heard about that today, from Bob Hampton, our quality assurance manager."

"Well, I wouldn't use the term so loosely, if I were you."

"What's that?"

"Calling Bob Hampton a quality assurance manager."

What Fred had said about Jason was on target. Jason was outspoken, forceful, and demanding. Jim knew he had his work cut out.

"I'd like to give you some of the details, if I may," requested Jim.

After a few words from Andrew setting the phone on fire, he settled down somewhat and listened as Jim related the details. They agreed Jim would supply people to go to the Camden facility to inspect all the goods on site, at Jim's expense, of course. Further, that Jim would fax a copy of a preliminary PIIF the next morning.

"Again, I'm sorry our first conversation had to be under these circumstances. I was honestly planning to call you tomorrow to say hello, and talk about my upcoming visit."

"Fine," was all Jason had to say before he abruptly hung up.

Jim immediately called Harold back to bring him up to date.

"How did it go with Andrew?" Harold asked.

"I don't know, but from what I've heard about him, he could have been tougher on me. Maybe he's giving me a little breathing room because he knows I inherited this problem."

"Could be" replied Harold, "but everyone in Aklin will see this as the last straw."

"I suppose so, but much of this is of their own making."

"How's that?" asked Harold.

"Well, you're aware that Aklin insisted on going to a new material, one our people repeatedly told them could cause a problem."

No," said Harold, "I wasn't aware."

"Come to think of it, you probably wouldn't be aware. There would be no way you could know about every engineering change that's made." Jim replied.

"That's not so, Jim. I put out a memorandum a year ago to all the plants instructing them to send me a monthly report noting every engineering change by major customer. I asked for that after we experienced some significant field problems, but I haven't seen the one you're referring to. If I had, I probably would have seriously challenged it."

"I see," said Jim, "I'll look into the matter and see what I can find out." After their conversation concluded, Jim asked Bob to drop by before he brought the rest of the group in.

"What's up?" Bob asked, taking a seat.

"I just spoke with Harold. He was unaware of the Aklin engineering change to the new material. He said he put out a mandate a year ago to receive monthly updates of all engineering changes."

Bob's forehead immediately turned a faint red and his facial features changed to a look of concern. Jim knew he had hit a sore spot.

"It's entirely my fault!" Bob admitted, pulling his shoulders back as if ready to take a bullet from a firing squad.

"I don't understand. What are you saying?" Jim asked.

"I stopped sending the Aklin end of that report over six months ago," he replied.

"Why?"

"Because Marion Brimmer insisted on it."

Who is Marion Brimmer?" asked Jim.

"He's my counterpart at Aklin, their quality assurance manager."

For the next few days, Jim and his crew attacked the problem with zeal. The result was they were able to stave off closing Aklin down, and kept them going with some extensive and expensive testing and inspection. In addition, Jim had been preparing for the Aklin meeting, scheduled in two days.

Overall, it had been a flurry of activity, calls and follow-up duties for Jim; and he was exhausted. At home, he thought about how things were transpiring. He had been forced to inform Harold that he had been receiving incomplete reporting for Aklin's engineering changes. Harold became upset and his initial reaction was to fire Bob, but Jim calmed Harold by explaining the full story.

Bob had had a number of conversations with Marion regarding the difficulty the plant was having with the timely implementation of Aklin's engineering changes. When Bob told Marion this was due to reporting and approval channels he had to follow, Marion insisted he do something about it. Bob said he discussed the matter with Nathan Carlton, the plant manager at the time, and his only direction from Carlton was do whatever was necessary to correct the problem. Carlton made it clear he wasn't interested in all the details; he only wanted his people to do the job. So, Bob decided to stop reporting Aklin data. Since, he reasoned, they could lose the Aklin business if something constructive wasn't done and it was the only solution he could see to speed up the process.

Jim explained to Harold that he had clearly let Bob know what he had done was wrong. Jim expected him and everyone else to abide by the rules, regardless of the situation. However, he did mention to Harold he gave Bob credit for having the fortitude to do what he thought was right under the circumstances. Although Bob's method was wrong, his motives were in the right place.

Harold agreed that Carlton had many good points, but he also had a reputation for being less than interested in details. Carlton didn't like being used as a sounding board for solutions. He saw any such effort as a sign that the person was either afraid or unable to make a decision, which was one of the reasons he was no longer at Denning. However, he made it clear that he didn't buy Bob's reasoning as acceptable and wanted to know what Jim planned to do. Jim told Harold he had already given Bob a written reprimand and was holding up his next planned salary increase, until Bob convinced him he had learned his lesson.

Harold agreed that the actions Jim had taken would suffice, for the time being. But he ended by saying that even one small slip from Bob, from this point forward, and he was going to insist on his job.

"You can't have a person in his position and not trust him to do the right thing. He's the quality assurance manager and the conscience of the operation," Harold remarked.

Things became more complicated. Phillip Brooks had received a call from the president of Aklin, who had learned of the problem and was insisting Brooks come and explain why Denning couldn't get its act together. Aklin had to issue a potential hold to its customers and was getting a lot of heat. Even though Aklin was clearly taking its current business elsewhere, it had considerable leverage with Brooks because of the potential for future contracts, where Denning had a leg up on the competition in cost and design.

Jim had to prepare for his meeting with Aklin, along with keeping things moving forward because of the unfortunate and untimely ALQA. In addition, the following day he had to meet with Brooks, Harold, and others, to discuss what Brooks would say in his meeting with Aklin's president.

The following day, the meeting with Brooks proceeded for what seemed like forever. They had been in his office since 8 A.M., and it was now only minutes until noon. Brooks had lunch brought in for all present, so they could continue the meeting without a formal break.

"So," remarked Jennifer Collins, Brooks' business manager, "you'll use the charts to show where we've made some substantial improvement in delivery and quality over the past few months and you'll speak of the organizational changes made."

"Correct," replied Brooks.

"All this is good, but if I were him, I'd be looking for that one thing we haven't put our fingers on yet," she added.

"And, what precisely is that?" asked Brooks.

"It's precisely what we haven't put our fingers on yet," she joked and everyone laughed. Jennifer had a sense of humor, yet Jim had found himself more impressed with her business savvy and her foresight. Obviously the reason she has the job, he thought.

"All right," said Brooks, "does anyone have any bright ideas?"

Jim had been quiet, providing input only when he was asked questions about the ALQA. When no one had a response, Jim slowly raised his hand. In doing so, he glanced at Harold, who looked surprised.

"Yes, Jim?" said Brooks.

"I'm scheduled to visit the Camden facility tomorrow and meet with Jason and his staff, and I will be proposing some unconventional actions we intend to take, much of which I haven't had time to discuss with Harold. Given the circumstances and our time crunch, I wouldn't feel right making the trip on the same day you're going to visit the president and telling them something you weren't aware of."

It was obvious Jim had Brooks' and everyone else's undivided attention.

"I didn't know about your visit, but it's great. So, what do you plan to offer them?"

"Well, again, I haven't as yet discussed my plan with Harold."

"Don't worry about that, Jim, just spit it out. We're all family here," said Harold, though Jim knew he was about to take a serious career risk.

"Yes," encouraged Jennifer, "that's why we asked you to attend."

"All right, here's what I was hoping to tell Andrew and his staff. Workmanship is by far the biggest quality problem. So, until we can start manufacturing practices and techniques that will make us a world-class manufacturing operation, we're going to insert a 200 percent dock audit procedure on all Aklin products."

Jim went on to explain the details of the procedure, which called for a 100 percent re-inspection of all products, for workmanship, after they were packaged, and another 100 percent check if rejects reached a prescribed level.

"I like it!" said Jennifer.

"Isn't that going to be expensive?" interjected Justin Pearl, the chief financial manager and a direct report to Brooks.

Jim glanced at Harold, waiting to see if he was going to jump in. He didn't.

"Yes, it will be expensive, but I don't see any other choice. The key is to get the kind of manufacturing practices quickly in place, so this isn't necessary. However, I believe it shows our intent to do whatever it takes to tackle and fully resolve the problem, while at the same time making sure the customer isn't penalized in the process," answered Jim.

"I have to agree with Jim," said Harold. Jim was relieved to know Harold was on his side.

"I also agree," remarked Brooks, then he continued. "Jennifer, I want you to get with Jim and work up a presentation on this, so we can take it with us. Justin, you'll also need to get a flavor for the cost of all this. I want him to know we're prepared to spare no reasonable expense in resolving our problems, once and for all." He stopped for a moment, then turned to look at Jim. "And Jim, thanks for bringing us up to date and for the excellent idea."

"You're more than welcome, but I might mention that a lot of the credit should go to Bob Hampton, our quality assurance manager. It was his basic idea."

"We'll make note of that for the record," said Brooks.

Later, Jim prepared to leave for his big day with Aklin. He was to arrive at noon, with the meeting planned for 3 P.M. As he was leaving, Linda let him know the president of the union, Mark Hudson, was outside and wanted to see him for a few minutes.

Phil Tanner had introduced Hudson to Jim earlier, but this was the first time they had met one on one.

"How can I help you?" asked Jim, after they had both taken a seat.

"I hear you're off to visit Aklin, to try to save the business. So, I just want to wish you good luck," said Hudson.

"Thanks, I'll do my best."

"But the main reason I'm here," he continued, "is we've got a problem to discuss."

"Oh? What's that?" Jim asked.

"It's probably best not to get into it right now. I know you're pressed for time. But when you get back, I'd like to discuss it with you."

"Have you spoken with Phil about it?" Jim asked.

"No," he replied bluntly. "I want to talk to you, not Phil."

"Well, Phil is our HR manager."

"Are you telling me that if I need to discuss anything with management that it's going to have to be with Phil?"

"No, of course not," replied Jim. "I'm just saying it should probably start with Phil."

"Well, thanks for nothing." Hudson started for the door, but Jim called him back.

"Mark. I'm sorry if I've upset you, but I am on a tight schedule today."

"Screw that. It doesn't have anything to do with the subject," he interjected heatedly. "You've made it clear if I have a problem I should take it to someone else. That's going to make it tough on both of us before all's said and done."

"Look, Mark. Don't assume what my intentions are. I only asked if you had discussed the problem with Phil. I didn't say we couldn't talk about it when I return."

"Forget it," said Hudson and with that turned and left Jim's office.

WARRING SCORECARD, PART 2

I stated that measuring a plant manager's effectiveness wasn't a matter of sizing up accomplishments verses failures. The truth is, it has more to do with how a plant manager reacts to the unexpected and unknown.

The Warring story this far shows you how often the unexpected and the unknown arise. It represents what most plant managers face daily, which is why many do not succeed in making a total success of the position. Handling the unexpected and unknown isn't something taught in college. As a result, managers are often left to their own intuitions as to how to respond and react, thus this books' expressed purpose.

As this ties to setting the stage for a successful waste-free manufacturing effort, other influences can and often do come to hinder progress. While we can espouse the benefits of lean manufacturing and hope it takes precedence over everything else, the truth is it usually never happens. While it can serve as the ultimate mission for manufacturing and help drive the basic concepts of production, the other aspects of running a business remain active after beginning such an initiative. The operation has to meet certain criteria established by management (e.g., profitability targets). The operation must also address what I call the *soft* aspects of the business.

The soft aspects have to do with the continuing task of reducing internal and external turmoil. This means keeping employee morale and the general respect of the community at an acceptable level. Most experts and consultants of lean manufacturing do not address this. As we will discover over the next decade, as we focus more activity on such principles and concepts, we will view training essential in handling the soft side of the business as imperative to any lean manufacturing effort. As this applies to the Warring scorecard, the following serves to note Jim's accomplishments and failures:

As Plant Manager

1. He used good methods to draw out the best thoughts and ideas of his people.
2. He gave his people credit for their ideas when the opportunity arose.

3. He handled himself well while interfacing with those outside the operation.
4. He stood up for his employees when it was necessary to do so.

Jim, however, made some additional errors in judgment:

1. A continuing failure to establish a good working relationship with the union.
2. A continuing failure to recognize and utilize the extended staff.
3. A poor reaction in his first meeting with the president of the union.

As Chief Conductor for Lean Engineering

A huge problem is starting to develop. Jim has been primarily focusing on daily production issues and has found little time to establish and implement a course of action that will serve to make the manufacturing end of the business more globally competitive. Further, he hasn't met the obligation to be a solid ambassador for driving such a process into the other facets of business (e.g., the enterprise as a whole).

Therefore, on being the chief conductor for lean engineering, we have little to comment on so far; which as you will see does little to establish a course that will clearly set the company above its competition.

4

Staying Focused

YOU HAVE PROBABLY SEEN ENOUGH EXAMPLES in the Warring saga already to see that staying focused as a plant manager can be difficult. Too many distractions draw one's attention away from the things he or she had hoped to accomplish as primary objectives.

Remaining focused on at least one driving conviction is of utmost importance. This serves to tell people you have a purpose in mind and a reason for being there, other than just handling daily problems. If a plant manager's only existence is seen as the latter, then he or she is usually perceived as little more than the head referee. This is why you must understand the importance of delegating authority.

Doing a good job of delegating authority involves understanding and practicing the following:

1. Deciding the things on which you are willing to concede full and absolute command, so others can manage without your interference.
2. Avoiding the temptation to take command of a situation occasionally where you have delegated authority.

Practicing this is not easy, and most plant managers do a bad job of it. I have had many managers tell me they believed strongly in delegating authority. What I often found was that they actually believed in selective delegation. In other words, they were willing to pass on duties that typically fell under their jurisdiction as managers but never the authority to make decisions without their expressed consent. While, in one sense, this may be delegating, it certainly is not delegation of authority.

The manager must be careful to avoid the temptation of reassuming authority. For example, let's say a plant manager has delegated authority for all factory scrap and rework matters. Things go well until a senior employee approaches the plant manager about an issue with scrap costs. Rather than tell this employee that he or she has an individual in charge of handling these matters for the plant, he or she makes a quick decision about what will be done. The result is the plant manager has usurped the authority of the person put in charge and has sent the wrong message to the workforce.

True delegation of authority means being willing to step back and allow others to assume command, at all times and without interference. Therefore, deciding on the things the plant manager is willing to delegate is important. This can vary greatly, depending on the situation. A key factor in all this is the depth of the challenge involved in carrying through those initiatives a plant manager views as critical to the success of the operation. In other words, if a plant is reasonably stable and running adequately, then the extent of delegation would normally be minimum. On the other hand, if the plant is experiencing major problems and a significant change in course is required, then the degree of delegation must increase proportionally.

There are a few things, however, a plant manager should never delegate to others, regardless of the circumstances. These include the following:

1. Review and approval of financial budgets and forecasts
2. Appropriate interface with customers and union leadership
3. Hiring, termination, and layoff decisions

Outside of these, it is up to the discretion of plant managers as to what they decide can and should be delegated to others. Let's examine these in more detail:

Under no circumstances should plant managers allow anyone the authority to submit a financial budget and/or forecast for the operation before they have had the chance to review and approve it personally. The staff and others must understand that any composite financial document for the operation (which generally consists of the combined projected expenses for all departments and operating units within the factory), must have the plant manager's expressed approval before it can be submitted. Never allow anything short of this to occur because if there is one area a plant manager can get into trouble the fastest, it's this area.

Plant managers must remember they are ultimately held responsible for the operation fully meeting all stated financial budgets and forecasts. As a result, plant managers must make certain they are *comfortable* with what is submitted to the powers above. When I say they should be comfortable, I do not mean absolutely certain they will meet the projected forecast. As with any good budgeting and forecasting, a reasonable challenge should be built in. Otherwise, it becomes a cover your fanny exercise.

In addition to budgets and forecasts, plant managers must make certain they have not delegated appropriate interfaces with both customers and union leadership entirely to others. Also, plant managers must be involved with hiring, termination, and layoff decisions. But, common sense and sound judgment, regarding the specific degree of personal involvement, is the best judge when it comes to both these matters. Just be certain your role in these remains reasonably active at all times.

HOW TO APPLY YOURSELF TO AN INITIATIVE YOU DON'T FULLY SUPPORT

Staying focused involves not only dealing with distractions in the form of daily problems but also in carrying out company initiatives you may not fully support, or for one reason or another, feel should be delayed for the time being.

This can especially become a problem in larger companies where personnel constantly seem to be pushing new and different programs. In my tenure as a plant manager, company-driven initiatives always seemed to get in the way of the primary objective I wanted to achieve. In some cases, I was less than supportive of the new direction. I don't say this to be critical. I say this to point out that new plant managers should prepare for just such an occurrence because, in all probability, it's going to happen.

Not surprising is that when a company is experiencing substantial problems, across a number of its operating units, activity in this area usually picks up. If only one or a small number of the operating units are having problems, the company is usually reluctant to change its established mode of operation. Unfortunately, most companies should be considering change during the good times, as vividly pointed out in *Fast Track to Waste-Free Manufacturing.*

Regardless, at times plant managers will have to carry out an initiative they do not fully support. How to apply yourself when this happens is important, for it can make a difference in your ability to stay focused on the things you want to achieve, while at the same time carrying out the special duties that go with the position. Here are some good things to remember regarding this matter:

- Accept that there are going to be some serious distractions you must face in seeing your primary mission through to completion.
- Accept that everyone will not see your objective as the most important issue facing the company.
- Unless the initiative seriously goes against the grain of your goals, support it fully and wholeheartedly. After all, upper management feels it is important.
- Remember, this is a great moment to utilize delegation of authority, if possible.

If the initiative seriously conflicts with the primary objective you've set forth as crucial, then prepare to put your job on the line. That's right, I said your job but not your career. As plant manager, your conviction about the primary objective should be strong enough that you are willing to stand up and fight for it. If your conviction isn't that strong, then it really isn't that important to you in the first place. So, just do what you're asked and be quiet about it. Don't complain or argue with others because, frankly, it's a moot point. Save your energy, do as you're instructed, and be seen as a good trooper. On the other hand, if your convictions are strong and you fight and lose the battle, then pack your bags and pursue your career elsewhere. Though it is unfortunate, it can and does happen.

What I find disturbing in this day and age are that people profess strong convictions but are unwilling to fight for those convictions. This isn't conviction in my judgment. It's simple rhetoric or, as my father used to say, *talking to hear your head rattle.* So, if you have little more than a head rattling conviction, it will not be long before your actions bear this out for one and all to see.

What I mean to suggest about true, unerring conviction is you make your primary mission something you truly believe. If it's not earth shaking, that's all right. It doesn't have to be. However, be sure you would be willing to put your job on the line for it.

One way of effectively managing an initiative you do not support is to give it away to someone. Here is the opportunity for you to delegate authority in a manner which shows support, yet it provides you the chance to refocus on the objective you feel critical to the operation's success. Care must be taken, however, because one of the things you don't want is for people in high places to see this as something you're delegating simply because you don't care.

You can demonstrate care and concern for the initiative by practicing good follow-up and reporting techniques, which serve to give it high visibility and exposure. For example, you can ask the person you've assigned authority to attend your daily business meeting and give a short update on progress. If reporting daily is too much, then make it weekly, bi-weekly, or even monthly. But the key, again, is to keep visibility and exposure at a high level. This demonstrates the importance you place on the initiative to others, both inside and outside your operation.

Of course, you should follow up with a series of scheduled meetings with the individual. Here the discussion should center around overall progress and what, if anything, you must do to help get the job done.

I mentioned this technique to a person who didn't think much of it because he felt my motives were less than pure. My response was that he couldn't be more wrong. My motive was to see the initiative through to a successful conclusion, and my actions clearly demonstrated that. I went on to say he could be confusing feelings with motives because a person can dislike an idea yet work in an active and responsible manner to make it happen. His response was typical of people's conviction about support: If you're not jumping up and down, clapping your hands and singing alleluia, then you can't possibly be behind my idea. Passive, yet constructive, support meant no support at all.

From time to time, plant managers are going to be faced with company-driven initiatives that seriously interfere with effectively implementing their objectives. As long as these initiatives do not directly conflict with those objectives, then simply develop a means of constructive support, which doesn't have to be an in-depth involvement.

THE WARRING ADVENTURE, PART 3

Jim's meeting with Aklin went well, overall. He outlined the plan for the new dock audit procedure and committed to have all workmanship problems fully resolved within 30 days. Even Carolyn Hamilton went easy on him and seemed quite impressed with the commitment he was making. The biggest surprise for Jim was how well he and Jason Andrew hit it off. Before the meeting, Jason took him on a tour of his factory and they seemed to share a lot in common about how a manufacturing operation should be run. Although Jason was rough cut and seemed to punctuate every sentence with a well-placed word or two (of the kind you didn't hear in church), he knew his business and he was a sharp, perceptive individual.

While they were alone in the plant, Jason freely offered Jim his opinion about various individuals at Denning. He disliked Bob Hampton (thought he was totally unqualified for the job.) He also didn't care for Fred Johnson, the sales and marketing manager (thought he was simply an overrated order taker.) He didn't respect Harold Jenkins, Jim's boss (thought he couldn't be trusted.) He absolutely despised Nathan Carlton, the previous plant manager (thought he was a total ass.) He did, however, like Zack Milen, the engineering resources manager and Joe Thompson, the production manager (thought Jim should have more like them around).

Jim decided to listen and avoid pursuing Jason's reasoning behind his feelings, at least at the time being, but he was in for an even bigger surprise. When Jim asked how he felt about Chuck Henson and Cindy Mabrey, Jason replied, "I like them. At least they don't give you a lot of BS. They tell it like it is, and their word can be trusted."

"Hm," pondered Jim. "The feedback I was getting was they kept the customer irritated."

"If that's coming from anyone at Aklin, it has to be Carolyn Hamilton," Jason responded.

"Carolyn Hamilton?"

"I'd say so. She's pushing some fast engineering changes and when she gets told something is impossible to implement, she can get very testy. Plus, being at corporate, she has the ear of all the top brass."

After the formal meeting, Jim accompanied Jason back to his office before departing. During the course of the conversation, Jason asked Jim how his working relationship was going with the union. Jim proceeded to explain that he had been so busy he really hadn't gotten to know the union leadership very well. Jason looked surprised.

"You've been there six weeks and you still haven't gotten to know them yet? I'd say you're about five weeks and four days short of where you should be at this time."

He related what he saw as the importance of a plant manager building a good working relationship with the union. He took pride in his relationship with the union and it had saved him countless headaches over the 24 years he had been a plant manager. His last words as they shook hands were, "Be sure to get to work on that union thing when you get back. All right?"

Jim assured him he would. On the trip home, Jim kept running things over in his mind. He had heard nothing but bad news about Carolyn and Zack, yet Jason said only good things about them. Jason also stated, for whatever reason, that he didn't trust Harold and Bob Hampton was in over his head. On the other hand, he had a great deal of respect for Joe Thompson and Zack. While Jim had his reservations about Joe, he made a mental note to get to know Zack better.

While Jim was smart enough not to overreact to anything Jason had said, he convinced himself to find out more as the opportunity arose. It was obvious that as rough around the edges as Jason appeared to be, he was no dummy and he knew how to be effective. Everywhere they went in the plant, Jason was greeted warmly and people responded almost affectionately to him. It didn't take a genius to see he was appreciated and respected by his people.

As a result, Jim took special note of what Jason had to say about the union. His thoughts then turned to how the conversation had gone with Mark Hudson, president of the union. If Jason were right, and Jim had no reason to believe he wasn't, then he had definitely gotten off on the wrong foot with the union. He decided he had to correct this quickly.

The next morning Jim asked Linda to set up a special meeting for him with Mark Hudson. According to Linda, Hudson reluctantly accepted the invitation, and when he arrived he was clearly not a happy trooper.

"Mark, I want to apologize for the way our meeting went the other day. I was wrong and I handled it very badly. So, I would like for you to give me another chance."

"Sure," replied Mark and though his expression had softened somewhat, it was obvious he wasn't buying in just yet.

"You mentioned we had a problem we needed to discuss. Is it too late to revisit that topic?"

"Could be," replied Mark.

"Well, let's test it and see," said Jim. After a long silence on Mark's part, Jim continued. "What's the problem?"

Mark shuffled in his chair, leaned back, and looked at Jim for a moment, as if deciding if he were indeed serious.

"All right. It's like this. Everyone wants to know if and when we're going to lose the Aklin business and, if we do, who's gonna be affected. I keep telling them I don't know. They look at me like I'm either lying to them or I'm just too stupid to understand what I'm being told."

"Stupid?" responded Jim. "What do you mean?"

"I mean stupid. They feel if I'm the president, then I must know something about it and if I've been told, then I'm either a yes man to the company or just too stupid to understand what you people are telling me."

Jim couldn't help chuckle a bit, but he cut if off sharply when he saw Mark was serious.

"I see," replied Jim. "Then what you're telling me is you need something to take back to them on this matter?"

"You got it," replied Mark. "Look, Mr. Warring, I'm not here to ask you to tell me something that's confidential between you guys and Aklin but, if you can share anything, it would help me. I'm telling you people are up in arms over this. They know you went to Aklin yesterday, but most everyone has written that off as a fruitless exercise, due to you being so new on the job. They've even told me Harold Jenkins said as much."

"Harold Jenkins told someone on the floor my trip to Aklin was going to be fruitless?"

"That's what's going around," Mark replied.

Jim could hardly contain himself. Harold had said almost the same thing to him. His mind immediately flashed back to what Jason Andrew had mentioned about not fully trusting Harold. This was starting to trouble Jim, more than just a little.

"I'm sure someone's taking what Harold had to say out of context. The truth is, I did meet with Aklin and the meeting went well. I will be honest with you, however. The chances of us salvaging this particular account isn't good. It's just too late in the ball game, so to speak, and it has nothing to do with me being new on the job. In fact, Jason Andrew and I hit it off. But, they've already been in negotiations with other suppliers and I'm certain they must have signed a contract with someone. The good news is they have a big new contract starting early next year, and we're going after Aklin's business on this."

"But you left people with the impression there was a chance to salvage the business," said Mark.

"That was before I knew what I know now," Jim responded.

"That's not going to sit well with most people," said Mark, looking troubled.

"But you said yourself that they didn't put a lot of stock in me saving the business."

"No, but they did like what you had to say in the meetings and you sort of inspired a lot of them. This isn't going to help that. I guess what I'm saying is while they're talking about you not being able to change Aklin's mind, inside they were hoping you could."

"I'm sorry. I wish I could have done something, but it was just impossible," said Jim.

"If you had told me all this before you left, I could have gotten the word out that our chances weren't good, but you were going anyway to get some new business for us," replied Mark.

Jim thought for a moment, while echoes of Jason Andrew's words drifted through his mind. He looked up at Mark, smiled, and said, "You know, you're right."

Later in their conversation, Jim suggested he would like to set up a weekly meeting with Mark to discuss general business issues, in order to keep him informed and up to date on matters. Mark thought it was a good idea but wanted to bring a member or two of his committee with him to the meetings. Jim assured him that would be fine.

Mark reminded Jim that contract negotiations were now less than nine months away and that each side was going to have to start some serious planning. He assured Jim no crucial issues were at stake but said that could change, especially if a big layoff occurred because of the Aklin issue. Maybe he should have known, but he was totally unaware of upcoming contract negotiations. However, he didn't mention it. He could just imagine how Mark would take it.

After Mark left, Jim reflected on a number of things. He was thinking he had now been in the position for six weeks and it seemed he had surprises every day, all of which were taking some considerable time away from the things he needed to focus on. The job wasn't what he had expected, but most troublesome was what he was starting to hear about Harold.

Although Jim reported directly to Harold, he had not been instrumental in recruiting and hiring Jim for the position. In fact, he had made it clear to Jim that he was Spurvin's choice for the job. Anthony Spurvin was Denning's corporate HR manager and had been extensively involved with recruiting and hiring him. From the beginning, Jim had sensed Harold wanted to keep some space between them. Jim didn't feel as close to Harold as he had with his previous boss and long-time mentor, Frank Zimmer. He told himself all this probably meant nothing.

He had not talked to Spurvin since coming aboard and had been told he was overseas handling labor negotiations for some small plants Denning ran in the European market, out of Ireland. However, when he returned, Jim hoped to discuss a few things with him. But for now, he felt all alone and somewhat unsure of what his next move should be.

He was jogged out of his thoughts when Linda said she had Patricia Owens on the line. Owens was the corporate QA guru and reported directly to Phillip Brooks.

"Jim," she said, "this is Patricia Owens. I'm calling because I need some of your time."

"Sure," he replied, "how can I help?"

"Can you drop over at headquarters around three?" she asked.

"I am rather busy today, but I can make the time," he replied.

Sitting now in Patricia Owens' plush office at headquarters, Jim was wondering if he hadn't missed his calling. Unlike his office at the plant, the atmosphere was quiet and relaxing, without the constant chatter of machines and forklifts running up and down the aisles.

"How have you been?" asked Owens.

"Oh, fine. And you?" he replied.

"Just great. Jim, I've asked you over because I've been charged with picking one of our plants to pilot a new program. She handed him a slick brochure that held the title DQS2000+.

"The acronym stands for Denning Quality System, year 2000 and beyond. It's kind of catchy, we think. Actually it's not a program but a process, and it incorporates our version of the automotive industry's QS9000 procedure. Have you heard of that?"

"Most definitely," Jim replied, "and I agree. It's a process rather than a program. In fact, it's extensive and will have an impact on every area of the business."

"Right. How do you feel about that?" she asked.

"To be perfectly frank, I have mixed feelings."

"What do you mean?" she requested.

"It's a little complicated, but what it boils down to is the conflict it imposes on getting anything else important done. It's simply a time-consuming initiative for everyone involved. I'm basing that comment on my experience with QS9000 and not knowing precisely what you have in mind for DQS2000."

"2000 Plus," she corrected him with a smile. "It's essentially the same process, but it's geared toward our industry's particular mode of operation."

"So, you need to pick a plant to pilot the process?" he interjected. "Were you thinking about my operation?"

"Yes, I was and here's why. With your plant near headquarters, so to speak, doing the pilot here makes sense. This way, I can work closely with you and your people to insure it's a full success. You know the importance of that, I'm sure."

"What if I said I thought the timing was poor?"

"In what respect?" she fired back.

"With respect to us having issues making the plant less than a favorable candidate. I've got a plant focus in mind that's going to take the full energy of everyone. I plan to incorporate a high level of lean manufacturing concepts and techniques, which will change almost everything we do. Doing this at the same time we're piloting a process of the magnitude you're suggesting would be next to impossible."

"Oh? I would think they could go hand in hand."

"Not as I see it," he replied.

"And why's that?" she challenged.

"With all due respect, I believe I just told you why."

Owens paused, then leaned forward for emphasis. "Jim, I want to be diplomatic about this, but you're making it difficult. I asked you over to seek your support, and I was hoping you'd see things my way. But, the truth of the matter is, the decision has already been made. We are going to pilot the process in your plant. That's a given. So, the only question left is when and how we start."

Jim sat for a moment studying her and thinking about what she had just said. An old friend had once advised that when he felt like exploding, pause long enough to count to ten. Jim took a deep breath, did the mental countdown, and proceeded.

"I think you've made yourself more than clear. I'm just disappointed that no one asked my opinion before it got cast in concrete," he said.

"To be truthful, someone should have. But it wouldn't have changed anything. You must understand that this sits right at the top of Mr. Brooks' priorities, and he's adamant about wanting to use the local plant for the pilot, which is your operation."

"Harold is also adamant, about me putting the things I just mentioned in place. Has anyone discussed this with him?" Jim asked.

"Certainly. He was in Brooks' staff meeting just yesterday where we went over everything in great detail. He fully supports this effort and simply asked that I bring you up to date on it."

Jim was thinking, *Strike three.* He had to work hard at containing his true feelings and saying something he shouldn't, especially in this setting.

"All right, I understand," he replied. "I'll do what I can to help."

"Great!" said Owens, slapping the palm of her hands softly on her desk and standing as an indication the meeting was over.

"I'll call you in the next few days and we'll start laying out a plan of action. All right?"

"Fine. I'll be looking forward to seeing you," said Jim and with that bid her a farewell.

When Jim arrived at his office, he walked past Linda without speaking and slammed his door with some authority. He sat down and placed his hands on the desk. He squared his shoulders, leaned back, and proceeded to breathe deeply for a few moments. Linda opened the door and stuck her head just far enough into his office to make eye contact.

"Is something wrong?" she asked, with a look of genuine concern.

"You could say that," he responded.

"Anything I can do?"

"No. Thanks anyway. I'm just cooling down a bit. Don't worry," he responded.

"Well. All right. If you change your mind, let me know," she offered.

"Thanks again," he said as she turned and closed the door behind her. After a while he picked up the phone and called Harold's secretary.

"Carol," he began, "Could you fit me into Harold's schedule sometime this afternoon?"

"I certainly can. How about now? Is that too soon for you?"

"No, it's fine," he said, "I'll be right over. Thanks."

"Hi Jim," Harold said. Jim sat down, but said nothing, waiting for a further gesture from Harold to proceed. "What can I help you with?"

"I just got back from Patricia Owens' office," he opened, then intentionally paused to wait for a reaction from Harold.

"And?" Harold finally asked.

Smooth response, thought Jim.

"She informed me that my plant was the pilot for the new DQS2000 Plus process."

"Right. I meant to discuss that with you, but I didn't think she would be getting to you this soon."

"I have to admit I'm just a little upset at this point," said Jim.

"About what?"

"About the way this whole thing has been handled. When I challenged her and tried to explain that timing just wasn't right for us at the moment, she told me, in no uncertain terms, the decision had already been made and I had nothing to say about it."

"That's a little overbearing, and I hope it didn't come down exactly that way. But the truth is, we have little say in the matter. I couldn't agree with you more about the timing being bad. But this is Brooks' baby and I have absolutely no control. On the positive side, however, it could bring some discipline to your operation. It's hard to criticize something that's proven so successful in other types of world-wide manufacturing."

"I understand all that, Harold, and I'm not taking issue with DQS2000 or QS9000 or whatever you choose to call it. My concern lies in the conflict this is going to pose in getting in place some of the things we've discussed as crucial to the operation. We simply—."

Harold raised a hand to quiet Jim. "Jim, we can sit here and debate this all afternoon, but it isn't going to change anything. I'd say the best thing for you to do is go back and think about how you get both things done. That's the key to all this."

"Any thoughts about how to do that?" he asked.

"That's why I hired you," responded Harold with a smile.

"Fine," replied Jim ruefully. "I get the picture."

In the heat of everything that had occurred, Jim had forgotten he had a meeting scheduled with Zack Milen, the engineering resources manager. He wanted to get to know Zack better, especially since he had heard so many favorable comments about him. Zack was sitting patiently in a chair next to Linda's desk as Jim arrived.

"I'm sorry, Zack," exclaimed Jim, "Something came up and our meeting just slipped my mind. I hope this hasn't posed an inconvenience for you."

"No problem," Zack responded. "In fact, I was just enjoying a pleasant chat with Linda," he continued, turning and giving her a broad smile. They went into his office where they spent the next 20 minutes getting to know each other better. In the process, Jim found out more about Zack's background and experience. At 32, Zack had managed three different industrial engineering (IE) and two manufacturing engineering (ME) departments before joining Denning, two years earlier. In addition, he had run a QA department for two years with an automotive company. At Denning, he was in charge of all the engineering resources, which included four industrial engineers, six manufacturing engineers and two tool engineers. As well, he also managed the tooling and maintenance departments for the plant. Overall, it was an important position, with a great amount of pressure. Everything Jim had heard and seen convinced him Zack was an outstanding manager.

The conversation shifted somewhat, as Jim tried to feel Zack out regarding his opinion about key players in the operation. Understanding what Jim was looking for, Zack was totally candid. In a nutshell, Zack liked Joe Thompson but felt he was from the old school of manufacturing and a bit too set in his ways. He highly admired Bob Frisman and felt he had very good potential. He thought Bob Hampton was a hard worker but sometimes questioned his judgment. He also liked Joe very much and felt he did a great job overall. Outside manufacturing, Zack said he highly respected Jason Andrew, the plant manager at Aklin, along with Patricia Owens, Denning's corporate QA manager. Jim couldn't let that one pass.

"Really? I met Patricia for the first time today," noted Jim.

"We've worked together on a number of problems and I feel like she has her act together. I admire her, but she can be a tough cookie sometimes."

"Yes, I found that out," replied Jim. An interesting thought came to his mind.

"You say you ran a QA department when you were in the automotive industry?"

"Yes, that's right," responded Zack.

"How did you like that type of work?"

"Very much," Zack added.

"If the opportunity arose would you consider doing it again?"

"Here?" Zack asked.

"Perhaps. Who knows? The opportunity could present itself and it's just good to know how people feel about such things," responded Jim.

"Of course. But since you asked, my ambition is to return to QA at some point in my career. That could be one of the reasons I think so highly of Patricia Owens. I understand enough to appreciate and respect the job she's done and what she's achieving."

"Interesting," noted Jim. "Have you heard about something called DQS2000 Plus?"

"Yes, she and I have discussed it in passing on a number of occasions."

"And what do you think about it?"

"I think it's needed, but I'm biased. I helped a company qualify for QS9000 certification, so I know the value of the process."

"What if I were to ask you to head up the DQS2000 Plus pilot program for us?"

Zack looked surprised. "I'd have to think about that. I mean, I'm not sure how Bob Hampton would react. I'm sure he would think he should be the person selected."

"What I'm talking about is more than just a conventional coordination role. I'm speaking about full and absolute responsibility for the pilot, with me playing a minor role. Bob is in his first management position and simply doesn't have the kind of experience you do. It's not anything we have to decide at this point. I'd just like you to think about it."

"Fair enough," replied Zack.

HELPING OTHERS MAKE THE BEST OF THEIR CAREERS

An important responsibility of the plant manager is to help others have a successful and productive career. This includes everything from giving advice and direction to helping steer individuals toward careers they may not have considered. It also includes correcting career positioning that simply does not fit an individual's professional qualifications.

How often have you seen people who were excellent individual contributors but failed miserably when promoted to management? What you often find are outstanding individuals being promoted as a means of swift remuneration. The feeling is, if they are not recognized and rewarded financially, they will leave for a higher paying job.

In many cases, the company would be better off increasing the individual's pay, in accordance with the promotion in mind, and leave them in the job they were doing. We all know it doesn't work that way. Normally, each position has established job grades and associated pay restrictions as it applies to equal pay for equal status. Still, it is unnecessary as it applies to pay in proportion to individual performance. As a result, people can and sometimes are thrust into positions that simply do not fit their professional forte.

Why a plant manager must emphasize this goes past just doing what is right for the individuals. It must be done for the overall good of the operation. If a company has even one or two clearly incapable people in key positions, immeasurable problems will occur.

The answer doesn't lie in firing such individuals. It lies in guiding them into positions where their talent and expertise can best be utilized and doing this in a positive manner.

How do you give a demotion a positive slant? It isn't easy, but things can be done to help. I have witnessed occasions without explanation as to why an employee had been pulled from a management or supervision role and placed back into an individual contributor position. The first assumption people normally make was the person couldn't do the job. Even if true, the focus should be shifted from that to the importance of the new role.

As an example, how often have you seen a single announcement addressing the person taking on a new assignment and the person being replaced? This usually involves some glowing remarks about the new person's qualifications and possibly a small statement about what the individual being replaced will be doing in the future, along with a less than sincere thank you.

An easier way to give the matter a more positive spin would be to prepare separate written announcements and, when possible, to post them on different dates. The first announcement would focus on the individual leaving the job and would highlight the person's new assignment and the importance of the change. Here, you can comment that a later announcement will follow, addressing who will be assuming the position. The second announcement would focus on the person filling the job vacated and, of course, their particular qualifications.

Outside of correcting position errors, as needed, the task then becomes to help others grow and get the best out of their careers. There are some key steps to doing this in an effective manner and it starts at the top of an organization:

Step 1: Know your direct reports extremely well. Take the time to understand their background (education, past experience, etc.) and their career aspirations. You might be surprised how many plant managers have never done this. I have known managers who have had people who reported to them for years and didn't know anything about them. That is, of course, outside of what might be included in their personnel records. Typical personnel records do not include, for example, that the employee would not relocate because of a daughter who has a serious, debilitating illness. Though company's records may have allowed the employee to request no relocation, he or she could have checked yes so as not be locked out of other future opportunities.

So, take care in using, as gospel, anything that is or is not in an individual's personnel record because it can mislead you. Even if the information is accurate, personnel records simply cannot and, therefore, do not contain the employee's full story.

You will often make rapid changes to fit conditions and circumstances as a plant manager. This means you and your people must be flexible, at all times. You cannot make the right decisions if you do not genuinely know the background, strengths, weaknesses, and aspirations of your direct reports.

Step 2: Insist that your direct reports do the same with their people. They should genuinely get to know their subordinates. Think of the effectiveness that comes to bear when both plant managers and their direct reports genuinely know their people. When an unusual condition arises needing someone in the organization to take on the special assignment, deciding on the proper choice is faster and easier.

Step 3: Establish challenging goals and objectives that meet the plant's primary focus. In many larger companies, individual goals and objectives (e.g., Manufacturing By Objective plans or MBOs) have become meaningless in the manufacturing arena. The MBOs have become burdened with things not related to the job people were hired to do in the first place. I am not saying MBOs are the problem (they aren't). I am saying that over the last decade they have become a victim of special interests.

However, with regard to MBOs, I do have some definite opinions. One of them is everyone should have no more than seven to ten major objectives. Any more than that and the individual will have difficulty staying focused. Another is to carefully weigh and measure each objective in relationship to its importance. Supposedly this always happens, but in my experience too little attention is paid to this matter.

Where this starts to go seriously wrong is when special interests, which have the ear of upper management, demand and get their share of all objectives. With too many special interests, fewer objectives will remain for focusing a plant on the things that would help it achieve world-class manufacturing status.

Believe me, many would dispute this. They would contend what they are pushing, as a key initiative that every employee should take stock in, is the one thing every plant must incorporate to become a true world-class operation. You know what? They could be right, but this doesn't change that in making the journey, a plant must do the right things first. A good comparison of what I am driving at is expecting a baby to run before it takes its first shaky steps. Read the following scenario to see what I mean:

1. The company decides to implement a new quality program across the board in all its manufacturing facilities. Those heading up the process demand and get approval for everyone to have at least one or two individual objectives pertaining to this initiative.
2. A corporate mandate surfaces for all plants to improve their safety records by a certain percentage. Here, those leading the process want everyone to have one or two safety-related objectives.

Get the picture? Before long, more than half of all individual objectives are consumed with initiatives that are driven down from others, outside manufacturing. Again, this doesn't make those initiatives any less important, but it boils down to a simple matter of priorities. If a plant is about to become an unfortunate business statistic, this is usually a sure sign it needs to apply all its energy to the basics.

Doing the right thing when it comes to this matter is, at best, like walking a tight rope. You can't afford to be viewed as some independent cuss who only has your own objectives in mind. On the other hand, you can't let others run rough shod over you. It's a balancing act. As much as possible, insure your people's objectives are in tune with the focus you've set.

Step 4: Conduct performance reviews every 90 days. Most companies have MBOs and most require a year-end review of achieved results. Doing this once or twice a year isn't enough to keep a proper emphasis. You should perform an in-depth performance review every 90 days or once every quarter. In doing so, you will discover the need to change or revise goals and objectives.

Most managers see performance reviews as an unwelcome and unpleasant task that goes with the job because, in most cases, they have not been trained to handle them properly. Usually, it is the one time each year they meet with subordinates, in a setting designed to address performance without it being driven by circumstance. This is typically when the manager can give someone a pat on the back (for some progress) or issue harsh criticism (for instances of failure). This is normally the worst time to make judgments and comments concerning performance. The manager can more easily address performance, or the lack thereof, in an informal setting.

Actually, the process of effective 90-day performance reviews is fairly simple if two rules are followed:

1. Absolutely no criticism or undue praise
2. Focus on the future and not the past

As a result, managers can look forward to performance reviews. Whether conducted once a year or once a quarter, they are not the place to criticize an individual or to reap undue praise. Rather, these should be conducted as fact-finding missions. In many instances, managers take performance reviews as their opportunity for what is called constructive criticism. Believe me, there is nothing constructive about it. At best, it leaves hard feelings and demoralizes. On the other hand, undue praise will either leave an employee with a false sense of accomplishment or it will be seen as an insincere compliment.

The key is to stay on course and provide a forum that centers on fact. Is the stated objective being accomplished? If so, fine. If not, discuss what needs to be done. In some cases, the answer will be to change the objective because conditions or circumstances have changed. In others, you can provide some help. Stay away from praise for a job the employee agreed to do in the first place. If a given objective is being surpassed, then focus on why. Was that objective understated, in terms of difficulty, or has the person devoted far too much time to it and not enough to others? When an individual has clearly gone beyond the call to duty, a well-deserved pat on the back is warranted, but this should be the exception rather than the rule.

Make a genuine effort to spend at least 50 percent of the review focusing on the future. Here the conversation should move away from company-related objectives to the person's aspirations and what can be done to achieve personal goals. Sometimes, this can mean providing the time to pursue special education and/or training. Regardless, center this portion of the review on what is best for the employee and forget how this may or may not fit with company needs. In fact, I have occasionally advised individuals to seek employment elsewhere. I did this when their personal goals and objectives were in direct conflict with the opportunities that were likely to develop within the company.

Jim and Bob Hampton busily reviewed the data from Aklin regarding quality levels on the latest shipments. Since instituting the dock audit procedure three weeks earlier, only one incident of poor workmanship had occurred, and Jim, Bob, and others had quickly addressed this. As a result, an employee on the second shift had received a written warning, along with a two-week suspension. They immediately went to a 200 percent audit, as they had stated they would if such conditions developed. Jim and Bob were elated the latest report indicated no workmanship or quality problems whatsoever. It was a first, as far as Bob could remember.

"Well, the dock audit appears to be working," said Jim.

"Without a doubt," replied Bob. "In fact, just before I came over, I got a call from Marion Brimmer, and he complimented us about the progress we've made."

"Good. I'm glad to hear that. Still, this is a costly proposition for us. I just got some figures from Dersmond indicating we could miss our forecast this month," Jim noted. Alice Dersmond was the plant's accounting manager responsible for coordinating all factory budgets and forecasts.

"The 200 percent audit we put in place was what did it," said Bob.

"Not entirely, but it certainly was a factor. What we need to do is find a way to be more efficient with this process," replied Jim. "Have you thought about asking Zack to help you on this?"

"No, not really," responded Bob, surprised by the suggestion.

"After all," continued Jim, "Zack heads up the IE department, which is why it's here. To help make our processes more efficient."

"All right," replied Bob. "I'll do that."

"Speaking of Zack, I've been thinking of having him coordinate DQS," Jim stated. "Before you respond to that let me tell you why. Zack has a background in quality and with everything you have going right now, I just don't see where you can give it the kind of attention it needs." Bob was obviously shocked at Jim's proposal, but he maintained his composure.

"If you think it's best," he replied.

"Bob, I'm not sure it's best, but it is an alternative for us. We've got to get DQS going since we've been selected as the pilot facility, but we also have to make sure we fully meet our commitment to Aklin. I don't see how you can do an adequate job with both at this particular time," said Jim.

"You're probably right," replied Bob, "but I would really like the chance to handle coordination of DQS."

"Well, we don't have to make any firm decisions now although I have to get back to Patricia on this soon. I just wanted to run it by you and give you my reasoning. So, think about it and we'll discuss it further."

Two days later, Jim welcomed Patricia Owens to his office and closed the door behind her. They were scheduled to spend time discussing DQS, as Jim had come to call it.

"I hear some very good things about your quality at Aklin," Patricia began.

"Really? From whom?" quizzed Jim.

"From a number of sources, actually, but most recently from Bob Hampton. He called me this morning to discuss some things and made mention of how it was going."

Jim was wondering what things she and Bob had discussed but decided not to ask. He was certain she would let him know, at some point. Something about Patricia's manner seriously bothered Jim, but he was determined to set his personal feelings aside.

"I see. We are making some progress, but we still have a long way to go to prove we can deliver error-free products consistently," remarked Jim.

"True, but you're making some decent progress," she insisted.

"Well, thank you, and we'll continue to press forward on the matter," he replied.

"I'm sure you will," she said, pausing briefly before pursuing another topic.

"Bob tells me you are considering Zack to head up coordination of DQS2000 Plus," she stated, waiting for a response from him.

Jim could feel a warmth spread across his forehead and he was sure he was showing that blush he got when he was seriously irritated about something. However, he tried to remain calm.

"I see," he replied, "Bob discussed the matter with you?"

"Of course," she replied. "Bob and I have no secrets. It has to be that way. As I hope you know, all the plant's QA managers report to me on a dotted-line basis and I encourage them to come to me about anything that might be troubling them whether it's a personal or quality matter."

Jim thought she was a self-centered snob, but he was more upset Bob had gone to her about something they agreed to discuss later. This was the second time Bob had seriously disappointed him and he decided, if he could help it, there wouldn't be a third. But, before he could get anything out, Patricia continued.

"I have to advise you that I cannot support putting anyone other than Bob in this role and, more important, it should be you that does the major coordination."

He couldn't contain himself any longer and interrupted.

"Wait a minute!" he said. "I'm willing to work with you on this matter, but I won't have you dictating who is going to do what, especially what I'm suppose to do."

Patricia was clearly offended by his remark and stood, glaring down at him from across his desk. She lifted her hand and pointed a finger directly at him.

"Now you listen," she started, her voice trembling a bit. "When it comes to DQS, I would remind you that I report directly to the president and he's given me full authority to run this process the way I see fit. I will have a say in who takes on the key roles and I do expect the plant managers to cooperate and participate fully."

Jim waited and said nothing. After a moment, Patricia lowered herself back into her seat. She was perched on the front edge of the chair, with her shoulders squared, her head back, and her eyes fixed on his, waiting for a response. He had to work hard at containing himself and counted to ten before responding.

"I think you've made yourself clear. It doesn't matter what I think on this matter. So, why don't we stop wasting time and you tell me what you want me to do, and I'll let you know if I can live with it?" She continued to stare at him a moment longer.

"I really hate it's come to this, but maybe that's the best way to approach it," she agreed.

"I think so," he said.

"All right, here's what I had in mind," she said, then stopped for a second. "Jim," she smiled, which he could see wasn't entirely legitimate. "It's obvious we both have some rather strong personalities and that they can rub each other the wrong way if we're not careful. So, let's start over and make sure we don't come away from this harboring any hard feelings."

"Fair enough."

"So, let me hear your thoughts about Zack," she stated.

Jim thought *She just sat there a moment ago and told me there was no way she could support anyone other than Bob in the role and now she's asking me to explain why Zack would be a good choice.* He thought about what to say before proceeding. He decided that if she were willing to open the door regarding Zack, he might as well put his best foot forward to convince her.

"I haven't made any firm decisions and I told Bob that, but I thought about Zack for three reasons. One is he has a background in quality and with the implementation of QS9000, which you may or may not be aware of. Second is he has more management experience than Bob and third, Bob has his hands full with Aklin and other quality issues."

"Yes, I'm fully aware of Zack's background, and I like Zack very much," she replied.

"Then why are you so dead set against him having the role?" he asked.

"Because Bob is the QA manager, and we simply can't have our QA managers playing anything but the lead role in DQS. You have to think about how it would look if they were excused from this, that is unless the plant manager took on that role."

"What are you driving at with all the talk about the plant manager's role?" he asked.

"I think the plant managers must step forward and assume full command on DQS. It has to be their first priority, above anything else, and the workforce must understand this. I've made my feelings clear to Phil Brooks and he supports my position 100 percent."

Jim felt like asking that if that were the case, why didn't Phil Brooks put out a directive stating that's the way it's going to be and save everyone a lot of headaches? But, he didn't. Jim realized he was deeply in an unpleasant situation and, if nothing else, had developed a strained working relationship with a key individual in the organization.

WHEN AND HOW TO BUCK THE SYSTEM

As I've mentioned, the job of a plant manager often hinges on how well one responds to the unexpected and the unknown. This is also applicable to any managerial position.

I can't stress strongly enough the need for a plant manager to stay focused on the primary objective set forth as critical for the operation and to avoid spending precious time and energy on distractions. Though there is a time to go to war, most important is when and where to wage the battle:

1. **Know that you're fully prepared for battle.** Do not strike up a fight unless you have a reasonable chance of winning. To wage a battle without the proper weapons and/or armor is a sure way to lose a war.
2. **Prepare a battle plan before striking.** What would have happened if the Allies in World War II had met the enemy without a plan of attack? They would have probably been defeated. This same idea applies to your battles in the business world. Always approach them with a well thought-out plan.
3. **Be certain you really need to fight this battle.** One of the worst things to do is to fight over something without real bearing on what you want to achieve. Doing so can often hinder progress and unduly strain working relationships.

Mindfully choosing when and where to buck the system is most important. Therefore, remember what the primary driver for this is. Go to war only if it is over something that is unquestionably going to conflict seriously with your key objectives. For anything else, regardless of how unimportant or trivial you may believe it to be, be fully supportive.

Just as important is how you should go about bucking the system: Remain totally professional at all times and keep your emotions in check. A calm, professional dissension on your part will most often be met with a respectful, if not always agreeable, reaction. Avoid ill-prepared and destructive verbal battles, which can aggravate some and alienate others.

When a proposal or plan meets the above criteria, pause and think the following: "This is a something I must seriously challenge. However, I will not allow my emotions to get in the way of a thoughtful, professional dissension." If you are going to challenge something that you feel is inappropriate, present your views as an honest concern rather than a defiant objection.

One of the best ways of doing this is to avoid an off the cuff response or statement. Bite your lip, so to speak, and take time to analyze the matter constructively before stating your case. Taking the following steps before definitely committing yourself helps tremendously in dealing with such matters:

Step 1: State a serious concern if you have one. Do it in a manner that makes the point but doesn't leave the impression you would in no way support the proposal. Follow up on this by noting you would like time to think about it. At all times, keep your reaction to the demand or proposal professional and respectful. However, avoid taking an absolutely hard and fast stance at this point.

Step 2: Leave and mindfully evaluate the proposal. Before deciding you will make a serious challenge, see if you can accommodate the proposal or demand without it seriously crippling your efforts crucial to the operation.

Step 3: Challenge the person and/or initiative professionally. In other words, don't just pick up the phone, or in an informal way, let the person know you can't support the proposed initiative. Instead, insist on a formal meeting with those of substantial influence. If you feel you cannot stand before such an audience and present your case, forget it and go along with the proposal since you are ill prepared for the challenge.

Step 4: Present your case properly if you proceed to a formal meeting. Be certain that, in your own way, you cover the following questions:

- What is your primary managerial goal and specifically what does this do (in your judgment) to enhance customer satisfaction and overall operational effectiveness?
- How will the initiative in question hinder the progress of your primary goal and/or the rewards that should be evident, as a result of fully and successfully achieving this objective?
- Is there a more appropriate time and place for the initiative you have chosen to challenge? Remember, you are not rejecting the initiative but are concerned about its proposed timing. If nothing else, you can always suggest that the initiative be rescheduled, as long as it doesn't appear indefinitely.

Choosing to buck the system formally is a serious matter. You should only do so if you are absolutely certain that if you don't, the primary mission you seek to accomplish is going to suffer a serious setback. There is a time for this and it can apply to programs, processes, or special initiatives already in place. If this is the case, use the same basic approach noted to challenge these.

Remember, however, that even with the best of reasoning and logic, your objection could go unheeded and you will be forced to comply or rebel. If you comply, you must decide how to support the initiative with the least amount of disruption to your objectives. If you rebel, you may need to update your résumé and seek employment elsewhere. But, avoid being a fence straddler. You either strongly believe in the challenge you have set forth or you don't. If you do, prepare yourself to move on. If you don't, then avoid a challenge to begin with.

One thing I am stressing with this is to avoid the temptation for petty griping and complaining. They serve no purpose other than to alienate people whose strong support you may need later. In essence, the matter of when and how to buck the system boils down to the following rules of procedure:

1. Before making a formal challenge, be willing to put your job on the line. This is unquestionably serious business.
2. Be certain that without such a challenge, you cannot adequately pursue your mission crucial to the operation's future success.
3. Never make off the cuff criticism to others, no matter how you feel about it. Always remember to maintain a high degree of professionalism in dealing with this issue and to keep your emotions in check.
4. Do your best job in presenting your logic for a challenge. Never approach a formal meeting without being totally and adequately prepared. Put your best foot forward, which includes challenging without being perceived as critical of the item's or initiative's importance. Make the point that your objection is not against the topic but rather the timing and/or the potential conflicts.

Ginger hurried with dinner and afterwards she and Jim sat on the couch in the den watching television. She had snuggled next to him as the show continued and he, in turn, wrapped a comforting arm around her shoulders and pulled her closer. They had been married four years and cared deeply for one another, but things had been so hectic lately that these kind of moments together had been the exception rather than the rule. He missed the time when life seemed simpler. He often didn't get home until well into the evening, and he hadn't had a single weekend free for them to enjoy together since he had taken the job.

Now, as they sat having a quiet and relaxing moment together, he didn't know why but he sensed there was something different about Ginger. She had been exceptionally clingy the last couple of days and he didn't know why.

"Honey," he began, rubbing her shoulder softly as he turned his head slightly toward her and peered down at the top of her head, which was nuzzled warmly into his neck, "dinner was great."

She looked up at him and smiled warmly. "Good, I'm glad you enjoyed it. How was your day?" she asked.

"Oh, the usual and you?"

She gently pushed back until she was sitting upright and looked squarely at him. She smiled again and ran the fingers of her left hand through her long, auburn hair, which had always been a nervous habit.

"Jim, I've got something to tell you," she said.

"Really? Well, don't keep me in suspense," he joked.

Her smile slowly faded and a look of concern spread across her features. Jim was suddenly beginning to worry. He thought *She's not about to tell me she has a serious ailment or worse, is she?*

Ginger reached for Jim's left hand and held it lightly before taking a deep breath, pulling her chin back and looking him straight in the eyes.

"I think you're going to be a father," she said quietly.

Jim was stunned and speechless. He continued to look at her for a long moment before smiling broadly and reaching to pull her close to him. Ginger uttered a nervous laugh of relief as they held each other warmly and took in the moment.

They spent the rest of the evening talking about the baby and their plans for the future. Jim was overjoyed and Ginger was happy and relieved. They had discussed the matter and agreed they were going to wait on parenthood for a few more years, until Jim had settled into his career and things were better financially. But, now that it was a reality, their excitement overwhelmed them and they forgot any concerns they may have had about starting a family.

The next day at work, Jim floated on a cloud. He barely noticed some of the little problems that normally frustrated him.

"Boss, you seem to be in an especially good mood today," Linda mentioned.

"I am," he replied, smiling warmly.

"Why?"

"Oh, nothing in particular," he lied. They had agreed they would keep the news to themselves until she had confirmed it with her doctor. But, Ginger was sure she was pregnant, and Jim felt the same way. She had this special glow about her.

"Well, whatever it is, I hope you shower with it every morning," Linda said. "I haven't seen you in such a good mood since the first day you took the job."

The rest of the day posed no serious problems. Jim had sat through a monthly staff meeting Harold conducted with all his direct reports, which consisted of fellow plant managers at other state-side locations. Overall, the manufacturing sector inside Denning was doing reasonably well, performing on schedule and to established budgets. Harold pointed out that the one hot spot was Jim's operation, which was facing labor negotiations in a few months and a major layoff with the demise of Aklin's business.

When the meeting concluded, Carl Armstrong, the Florida plant manager, asked Jim if he wouldn't mind showing him some more of his plant. Jim was happy to accommodate and after a plant tour that lasted roughly an hour, they were sitting in Jim's office.

"I suppose you've heard the rumor," said Carl.

"There are so many flying around this place. Which one?" Jim joked.

"About Harold," Carl replied.

"No, I haven't. What's up?"

Carl was Harold's most seasoned plant manager, having run the operation in Florida for six years and was generally viewed by others as the person who would eventually succeed Harold. Jim and Carl had developed a good working relationship with one another and had a mutual level of respect, thus becoming relatively close. On two occasions, Carl brought his wife, Joann, providing Jim and Ginger the chance to take them to dinner and get to know them better. Ginger and Joann hit it off well, so Jim's overall relationship with Carl had grown past being just a business acquaintance.

"You're kidding, of course," replied Carl.

"No. I haven't heard the first thing regarding Harold. Though you may be surprised, I'm generally the last to know anything around here," Jim said. Although he seemed to be joking, he was quite serious about it. For some reason, everyone seemed to know what was coming long before he did. He reflected that it was a good sign he had not done his homework well in setting up a decent feedback loop for the rumor mill.

"Well," started Carl, "the rumor is Harold is headed for greener pastures."

"Really? What's that all about?"

"I don't know for sure, but I guess he wants a head position with another company. He's never made any bones about wanting to run a company and feels he's ready for it. I've also heard he and Brooks have had problems and Brooks could be pushing him to leave."

"If that happens, then you'll probably replace him," Jim supported.

"Not necessarily," challenged Carl. "I've heard Brooks is thinking about completely eliminating Harold's position and having the plant managers report directly to him. Even worse, I heard he's given some serious thought to moving Patricia Owens into manufacturing. If Harold left that would be a perfect opportunity for him to do so, since I don't believe he wants all the plant managers as direct reports. I'm certain that wasn't his idea to begin with but something one of his chief assistants has been pushing him to do."

"You're kidding?" noted Jim. "Do the other plant managers know?"

"I'd say so though I haven't had a conversation with any of them," Carl replied.

"If that happened and Patricia did take his place, I'd have to make a serious career decision."

"I think most of the plant managers feel the same way," replied Carl. "She's smart, but she walks around with blinders on. She's a my way or the highway individual."

"Oh, you've noticed that, too? I've been having a running battle with her lately over who's going to head up DQS2000 Plus. She wants to pilot the program in my plant and she's insisting Bob Hampton, my QA manager, be the coordinator. But I've told her I don't think he's ready for that."

"I'd have to agree with you," said Carl.

"At least that's good to know," responded Jim. "I was thinking about Zack Milen, my IE manager. Do you know him?"

"I've heard of him but we haven't met. If it's any consolation, what I've heard has been positive."

"Again, something good to know," replied Jim. "I'm impressed with the guy myself, and he would be perfect for the job. Patricia insists she's going to call the shots. Says she has Brooks' full support and backing, so I may have to use Bob whether I like it or not."

"Hey, ol' buddy," said Carl as he stood. "I'd like to continue this but I have to run. I have a plane to catch. It's good to see you again." Jim shook his hand warmly and bid him a fond farewell. He was looking forward to the end of the day himself and heading home to see how Ginger was doing.

WARRING SCORECARD, PART 3

As Plant Manager

A fair balance, overall. On the positive side, Jim listened much better to what others were saying and reacted accordingly. As a result of his conversation with Jason Andrew, who had much more practical experience, Jim held a constructive meeting with the union president and achieved an agreed-upon approach to better overall communications. One of a number of small failures in listening carefully, however, occurred when the union president told him contract negotiations were only nine months away. He should have immediately gone to Harold Jenkins to discuss why he had to find out from the union president and what was being done to prepare for the negotiations.

On the negative side, Jim chose to strike up a battle without being fully prepared and armed. Most serious was the conflict set up with Patricia Owens who was potentially at the top of the list as Harold's replacement, who was considering a career move as rumor had it.

Burning bridges along the way can often be a dangerous practice for the plant manager. In today's business world, the person one selects to fight today could be his or her boss tomorrow. This isn't to say a plant manager should be afraid to challenge others. However, some realistic thought has to be given to what a challenge will accomplish other than perhaps a level of conflict that could hinder a decent working relationship.

On the issue of listening carefully and responding accordingly, Jim failed to do this when Jason Andrew went out of his way to voice a serious concern about Bob Hampton, Jim's QA manager. Before this, Jim had adequate evidence Bob was slack in his duties. Though he shouldn't overreact to Jason Andrew's words, he should have realized a definite problem existed and that he needed to formulate a plan to put Bob on the right course or remove him from his present position. If there was a battle he should have waged with Patricia Owens, it should have been about Bob's very apparent shortcomings as the plant's QA manager rather than the implementation of DQS2000 Plus. As you will see later, this will prove to be an error in judgment with serious consequences.

As Chief Conductor for Lean Engineering

Outside of mildly challenging Owens with regard to his intentions of incorporating lean manufacturing and the potential negative impact of her DQS2000 initiative, Jim hasn't yet gotten off dead center. This is unfortunate in that he has a good background in the process and understands its value. However, like many young professionals in their first role as plant manager, he is finding it difficult to do little more than react to the job's daily occurrences. This, as you will see, will come back to haunt Jim.

5

Getting Effective Results

WHEN IT IS ALL SAID AND DONE, achieving satisfactory results is the name of the game. Even the most popular of personalities have to deliver at some point and for those who (for whatever the reason) do not possess the gravitating personality, effective results can often overcome what could otherwise be perceived as serious shortcomings.

Some leaders have the ability to inspire people to great heights in support of the mission they have outlined. Others don't because the great majority of managers are not Franklin Roosevelt or Martin Luther King, Jr., when it comes to sparking others into action.

The true measure of effective results is hard to define, in terms of a formula applicable to every situation. However, it can be summed up as the accomplishment of objectives that make an operation more competitive. In every case, this must start with a strong personal commitment on the part of the plant manager.

As an example, if the managers' principal commitment is to keep their job, then their primary objective is going to be to please those who hold power over them. On the other hand, if the goal is a higher mission—such as being unrelenting in moving the operation to a much higher level—then the primary objective goes far beyond personal consequences. Of course, fence sitters are everywhere. They usually hold enough conviction about something to get the ball rolling but then retreat at the first sign of any solid resistance.

Getting effective results starts with fully understanding what is expected of you and what you desire to accomplish. This isn't always thought out well by new managers because initially being placed in such a position is overwhelming. For the first time, they have people coming to them for advice and counsel. As a result, they feel as if they are going down the first steep incline on a roller coaster. The new job takes their breath away. Everything else is forgotten for the moment, other than experiencing the experience. Therefore, some of the best advice I can give the new manager is to put the newness of the experience behind you as quickly as possible and focus on what you need (and/or hope) to accomplish. Pursuant to this, keep a few things in mind:

- You do not own the operation or function over which you have been chosen to manage. You are the caretaker and are expected to leave it in better condition than it was when you entered the picture.
- Remember the expectations others have of you. This includes those who have an influence over your career and those who look to you for guidance, leadership, and direction. Sometimes in the heat of battle, it is easy to forget this and go with your instincts, which can be self-centered in nature.
- As much as you can, apply an emotion-free approach in your decision making. In other words, put aside personal feelings and look at each situation (needing your determination) as clearly as possible. One way to do this is to think about a judge who has well-established laws to guide the decision-making process but who, in the end, tends to tailor the ruling based on the particular conditions and circumstances. Often this decision will include a thorough review of all the facts. If you see emotion ruling the moment, regardless of the pressure, put off a decision until you can analyze the matter in a clear and unbiased fashion.
- Do unto others as you would have them do unto you. Many of the decisions and actions you will make will sometimes have a lasting impact on the future, livelihood, and/or career of others. As a result, many of your decisions have some potential significance, as applied to the effect they have on others. However, do not confuse doing the right thing pertaining to your position with an overzealous level of attention to the feelings of others. Given the conditions, keep in mind how you would like to be treated if you were on the receiving end of your decision. Often this involves a simple dose of common courtesy and personal consideration.

- Stay focused. I have found that, in a high-exposure leadership position, too many things can serve as a distraction. It is good, at least once a day, to sit back, take a deep breath, and remember what it is you initially outlined as crucial to the operation's success. Ask yourself: Is this in the best interest of that prescribed mission. If not, get back to it as quickly as possible.
- Keep an "Unresolved Issues" log. Develop this and refer to it often. List the critical issues and resolve each of these completely and thoroughly. Once done, delete and discard them from your list. Jim Warring has a number of issues he has not adequately addressed. One of these was brought to his attention early on and concerned Bob Frisman's inability to satisfactorily deal with financial matters. As you will see, this will come back to haunt him. Also, Jim has allowed other issues to fade from memory and/or active attention on his part. Of course, the obvious question pertaining to keeping such a log could be: How do you gauge what is critical and what isn't? The answer is simple. If there is any doubt whatsoever, include it in your log. If you do this, it won't be long before you will phase those things out that do not apply.

KEEP AN UNCLUTTERED MIND

In order to get effective results as a manager, keep your mind as fresh and uncluttered as possible. Just like your body, which needs a regular dose of rest and relaxation in order to remain healthy, your mind requires the same. But in today's business world, the common practice is to stress both mind and body with exhausting work schedules. The body will eventually rebel and force a person to stop, if for nothing more than a brief rest. However, the mind can still be churning away at a breakneck pace, dreaming about work or, even worse, having nightmares about it.

This is why I suggest occasionally taking a *day break*. That, of course, is a full day away from work on a somewhat regular basis. Some professional advisors hate this suggestion and would preach workaholism is the road to success. They would say taking a day off would send the wrong message to the troops, and the manager should be the first to demonstrate a work ethic that involves physically being on the job as much as possible. I adamantly disagree because my own experiences have proven this wrong. My worst decisions as a manager came when I was clearly approaching mental burnout. Conversely, my best efforts came after I had enjoyed a mental refresher from work, such as after a vacation or a holiday break.

This is one of the reasons I believe shorter, yet more frequent mini-vacations have become a popular practice. Managers and others in high-stress roles recognize they need a mental break on a reasonably consistent basis.

If I could set up the ideal scenario, it would be to take a one-day break once a month. Ideally, this would be on a Wednesday, right in the middle of the standard workweek. In fact, if I owned my own company, I would insist my key managers do this. I can assure you I would have managers who were sharper in their decision making than the competition taking a more conventional approach.

Before you point out that having every manager off on Wednesdays is unrealistic, remember I called this an ideal situation, not necessarily a workable one. However, where there's a will there's a way, and appropriate scheduling could accommodate the practice.

As this applies to the plant manager, work hard at forgetting how others may feel about it. I know when I first started this practice, I worried about how others would think and feared something terrible might occur that required my presence. I discovered that those left in charge were more than capable of effectively dealing with most issues. Otherwise, I made sure they had a number where I could be reached and was only a phone call away. As the manager in charge, you will be rated on how successful the operation is doing overall and not on how much time you physically spend on the job.

Most plant managers seldom take all the vacation they are due. I contend they should use every day they have coming, without exception. The key is to practice a common sense approach. Do not schedule your one-day break on those months with extended holidays (consisting of a couple days or more) or during the same month you are planning a regular vacation. Using this technique, you can end up having a refresher each month while not being perceived as spending too much time away from work. Regardless, keep your mind uncluttered by allowing it to refresh regularly. This isn't something you can satisfactorily achieve with an occasional coffee break at work.

LEARN TO DELEGATE EFFECTIVELY

As previously mentioned, keeping a precise focus in mind is the key to getting sound results. To do this, I practiced what I called a Four-Corner Technique.

Using this simple, yet effective approach starts by taking a blank piece of paper and drawing a square. On the outside corners of this square, note the four most important initiatives you need to accomplish in order to achieve the basic objective(s) you have set as critical to the operation's future success. In the middle of the square write the words *All other matters*. See Figure 5.1.

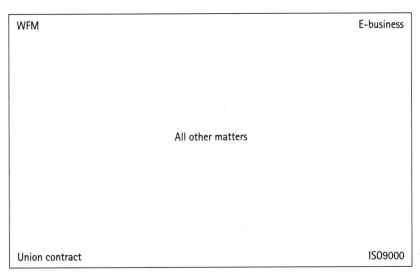

Figure 5-1. Sample "Four-Corner" Technique

Pin this up on the wall of your office or where you will see it every day. In doing this, you've laid the foundation for keeping your priorities straight and things moving in the right direction.

After completing this exercise, start by focusing more time on the initiatives outlined on the corners than anything else. Jot down, at the end of each workday, an estimate of how much of your time was spent on these initiatives.

What you may discover is that in the fast-paced and problematic world of management, you aren't spending enough time needed to see these through to a successful conclusion. If so, you must then decide how you will direct more of your energy to these initiatives or realistically face the fact that you simply don't have the time and/or fortitude to make them a reality. If it's the former, move on and make it happen. If it's the latter, it may be symptomatic of other problems.

It has been my experience that when managers decide the latter is the case, they haven't done an effective job of delegating authority. Things in the *All other matters* category were probably around before you assumed the job. As a result, you have allowed yourself to be sucked into a mire of old, persistent problems and difficulties that will continue to plague your operation until you undertake and implement the goals you have outlined as critical. This is, of course, assuming you are on target regarding what you need to accomplish.

If this is the case, then it's time for one last exercise. Note the specific issues at hand that fit into the all other matters category and that have served to take a considerable amount of your personal time and energy. As an example, it could go something like this:

- Consistent delivery problems
- Serious in-process quality problems
- Ongoing union and/or personnel problems

Provide yourself with more time to address the important mission. For every one of the noted issues, identify a *champion* to whom you will delegate authority. This champion would be responsible as your chief liaison in managing these business issues. However, in setting this up, you should have rules of engagement to pass on to these champions, in writing. Here are the rules:

- They must keep you informed of developing issues and their subsequent decisions.
- They are free to act on their own with the exception of certain circumstances (which should be clearly documented.) When such circumstances arise, they are to see you immediately for further directions.
- They must provide you with a monthly status report on the issues addressed and actions taken.

Regarding appropriately and effectively delegating, I have noticed a serious flaw in some plant managers, which is the tendency to micromanage the work of others. As a result, those delegated authority could be less concerned about taking appropriate action than satisfying the endless scrutiny of their supervisor.

If you are a micromanager, what I am proposing simply wouldn't be an effective tool. I say this because if you are not willing to let go of something, then delegation becomes a fruitless exercise and a completely despairing venture to everyone else involved.

This isn't to say that managers shouldn't stay informed or that they would never be required to make the final call. What I do say, however, is if you pursue detail over substance, then delegation can become more of a problem than a benefit.

THE WARRING ADVENTURE, PART 4

Jim was on the phone with Alice Dersmond, the accounting manager. He had just returned to work with Harold, who had accepted Jim's invitation for lunch in order to further discuss DQS2000+ coordination. Overall, it had been uneventful and Jim realized Harold would not support him in any manner that might be viewed by the powers above as being uncooperative or unsupportive on the matter. No matter how Jim felt, Bob Hampton would have to take on the coordination role, although again, Harold reinforced the decision was Jim's to make.

It was approaching month's end and, as he always did, Jim had called to get an update from Alice pertaining to how they were doing against their financial forecast for the month.

"It appears we have a problem," Alice stated.

"What's that?" asked Jim.

"Bob Frisman's scrap costs for the month have soared recently, and it appears he's going to be almost $30,000 over forecast."

"How can that be?" Jim responded.

"We're checking that now and probably should know something before the day is out. Based on what I see, it appears he's been scrapping some old, obsolete parts that have been on the books for some time."

"Well, if they're obsolete, I would think we should be making some appropriate disposition on them."

"The problem is we didn't forecast it and just as important is why we've built up such an obsolete inventory level in the first place," replied Alice.

"I'd like the answer to that myself," noted Jim, "but if it was on the books you or someone would have known about it. I'm wondering why I'm just now finding out about it?"

She was silent for a long moment. "Mr. Warring," she began, her voice hardening somewhat, "we do send you an obsolescence report every week."

"I'm totally familiar with that, Alice, but I don't have the time to wade through it every week and pick and choose what is or isn't an issue. I expect you to do that and keep me informed," he replied, becoming irritated with how the conversation was proceeding. "Look," Jim said, being more constructive, "get the facts as soon as you can and see me about it. All right?"

"All right," she replied and concluded the conversation with a stiff goodbye.

Jim sat for a moment and pondered what he just heard. He knew if there was one thing that could get a plant manager in hot water, especially under Phillip Brooks' regime, it was to miss the mark grossly on financial forecasts. In fact, Brooks had a practice of requiring any plant manager who experienced a serious forecasting problem to meet directly with him, Justin Pearl (the chief financial officer), and select others, to explain why. Jim had heard these meetings weren't especially a pleasant get-together. Brooks' well-known position was that though annual budgets might be difficult to meet precisely, due to unforeseen circumstances, monthly forecasts were so short range in nature that managers should have no excuse for missing one. And, what was now concerning Jim was that a $30,000 miss wouldn't be viewed as small change.

He knew he had to get to the bottom of this quickly and figure out a way to meet the forecast in spite of this unexpected drawback. He asked Linda to contact Joe Thompson, Alice Dersmond, and Bob Frisman and have them immediately come to his office. Some 15 minutes later, they were sitting across from him at his conference table.

"I've asked all of you here because I just received notification from Alice that we could miss the month's financial forecast in Bob's area by as much as $30,000," noted Jim.

He paused at this point to study their response. It obviously wasn't a total surprise to Joe and Bob, as far as Jim could detect. It was Joe who decided to speak up on the matter.

"I just got word of it from Alice about ten minutes ago," said Joe.

"Then you weren't aware of it?" asked Jim.

"Well, not entirely," replied Joe.

"Joe, you and I have discussed this on at least three different occasions," Alice interjected harshly.

Jim thought: *Well, what is it Joe? Just got word or knew about it for some time?* He decided the first thing he had to do was to remove the tension that was obviously developing and then proceed constructively, considering the depth of the problem.

"All right. Hold on. Let's see if we can get this down to something we can get our hands around and go from there," he commanded. "We're not here to place blame or fault. Ultimately, anything along those lines is going to rest squarely on my shoulders one way or the other. What I would like to discuss is how we got where we are and, most important, what we can do to correct an unfortunate situation."

Jim could see the tension on everyone's face relax a bit. Clearly, Bob wanted to say something, but he remained silent, glancing at Joe on occasion as if looking for permission to speak up. Jim decided to open the door.

"Bob, since this is your area, what can you tell us about the problem?" he pursued.

"It's like this," he started. "I'm not sure if you're aware, but we build sub-assemblies in our area for the Aklin account. As their business started to decline we've found ourselves with an overabundance of inventory that was a result of a volume buy for much higher levels of production. Joe and I discussed how to cut our inventory levels, per the instructions you issued, so we decided to dump the parts; since we can't send them back to the supplier. With the loss of the Aklin account we'll never get the chance to use them. As I'm sure Joe can point out, there was a misunderstanding on how to go about doing this."

Jim was starting to find it all very interesting. Bob was telling him that a significant obsolescence issue, perhaps created because of poor purchasing practices, had been around for some time. Though he, Joe, Alice, and others were aware of it, they had found it either unimportant or unnecessary to bring this to his attention.

"In response to what Bob has mentioned," Joe began, "I wanted to bring the inventory level down by forecasting and scraping a small amount each month until we had the matter behind us."

"Well, I'd be remiss if I didn't say there's more than one surprise in all this. First, I was not aware we built sub-assemblies for Aklin in your area, Bob. I'd like to find out more about why because producing something in P&S for the commercial side of the business seems strange to me. Further, while I don't expect any of you to confer with me before making daily decisions for your area, I am surprised this particular matter wasn't brought to my attention. Now, having said all that, let's focus for the moment, if we can, on where we go from here. That's really the big question we have to address at the moment."

"We only have eight days left in the fiscal month," noted Alice. "Therefore, it's going to be hard to do anything that is going to have an impact on the bottom line."

"We have to try," Jim responded. "I can't go to Harold about the problem without some sort of resolution on the matter. Let me ask you, Linda, is there any knowledge of this problem up the ladder? In other words, does Pearl know?"

"Of course," she replied in a matter-of-fact manner.

"Oh? And how does he know?" asked Jim.

"I told him," she replied.

"You told him?" he responded, surprised at her response.

"Certainly. I report to him on a dotted-line basis, as all the plant's accounting managers do, and I have a responsibility to keep him informed when it comes to financial matters."

Jim had to bite his tongue. He could feel the emotion rise because of her response and wanted to challenge it. He wanted to say: *Sure you have a responsibility to keep him informed, but not before everyone agrees it's a problem that can't be overcome. As a member of my staff, you also have a responsibility to this plant's reputation and you should take care before doing anything that would unnecessarily put it in a bad light.* However, once again, he paused long enough to decide this wasn't the appropriate time or place.

"You and I will discuss that further at a later time, Alice," he had to say before moving on, wanting to leave the door open on the matter. "But, for now, do any of you see another means of cutting costs in other areas to off-set this?"

"Well, we do have other planned obsolescence for Aklin that could be re-scheduled for next month, if we decided we wanted to do that," noted Joe.

"That sounds like a possible solution. How about it Alice?" Jim noted.

"If you like, I'm sure that would offset the problem this month. But, it still doesn't change the fact that our most recently developed forecast for the balance of the year is still going to be off by $30,000 or so."

"True, but we would have some time to adjust and see where we could make up the difference with some other potential cost-cutting measures," replied Jim.

Over the next 15 minutes, they came to the conclusion that re-scheduling some planned obsolescence for Aklin would be the best approach and with that Jim ended the meeting. After they had departed, he took the time to assess what had surfaced.

First and foremost, it was obvious Jim apparently had a problem with internal communications. Second, he remembered he had been warned early on by Justin Pearl, that Bob Frisman wasn't the best when it came to the financial side of the equation. It was something he should have fully addressed with Joe Thompson at that time. If there was any fault to be placed, it had to be on himself. Third, he was unaware that work for the commercial sector was being conducted in P&S, so he still had to work on fully under-standing his own business. Last, but not least, was Alice's attitude concerning where her chief obligations rested when it came to communications within structured reporting channels.

Jim thought about how to resolve the matter. He decided to start by hold-ing a special meeting with all his entire staff to address internal awareness and communications. He was convinced a discussion was needed to resolve when matters should be brought to his attention, along with when to com-municate with peers and associates, and how to be more effective in tracking and meeting financial forecasts. He wasn't sure about the particular format he would use in achieving this, but he was certain he had to do it quickly.

Jim decided to hold his session with the staff at a local resort, on a Friday, and extend an invitation to them to stay over for a Saturday golf outing or other activities. He even invited them to bring their spouses, since it would allow everyone a chance to have dinner together on Friday evening and to extend their stay over the entire weekend if they liked.

On Friday, they conducted a roundtable on the issues. Though at times the discussion became emotionally charged, Jim kept them on track and steered everyone toward a consensus resolution. He, of course, made some mental notes in the process.

On the positive side, his staff displayed the ability to work through tough problems, in a most effective manner. He was proud of that and shared his observation at the end of the day. Even though he left the meeting in his office with an unfavorable impression of Alice (as a result of her stiff posi-tion on reporting responsibilities), he realized she was one of the brighter and most underutilized talents on his staff. In this particular session, she had served to be a wealth of knowledge, input, and solid suggestions for improvement.

On the negative side, he wasn't impressed with Joe Thompson's depth of involvement throughout the entire session. He seemed preoccupied and at times totally disinterested. Jim was especially disheartened, considering the position Joe held. Also, Jim discovered that Bob Hampton seemed to gauge almost everything in terms of political posturing. He often swayed dramati-cally with his expressed opinions toward the most popular suggestion at that time. Though it could be chalked up as little more than an unappealing per-sonal characteristic, Jim was concerned about his QA manager. He couldn't help wonder if Bob acted in the same manner when it came to quality issues. If so, that could spell some serious trouble for the plant. He decided he would put some time and energy in finding out more.

The dinner also went exceptionally well. It was the first time Ginger had the opportunity to meet Jim's staff and their respective spouses. It proved to be something she enjoyed and it seemed apparent that everyone enjoyed her company, as well. That certainly came as no surprise to Jim, since she was one of the more friendly and outgoing people he had ever known. In fact, he had remarked to others, on more than one occasion, that it seemed she never met a stranger. The rest of the weekend went well, and on their way home Ginger remarked she thought everyone had a great time. Jim, as well, had enjoyed the session and was certain it was going to pay some important dividends. As he was about to discover, he couldn't have been more right.

"How does it look?" Jim asked Alice. She was in to give him an end of the month update against the forecast.

"We should make it. With the delay we put on the Aklin obsolescence until the last quarter and the other cost savings measures we came up with at the meeting, we could in fact slightly beat our forecast for the month," she replied.

"That's good," noted Jim. "Just shows you what can happen when you get the right minds to work on a problem. But changing the subject somewhat, what have you passed on to Pearl about the matter?"

"Just that we still have about $20,000 to make up in order to meet our previous forecast for the balance of the year. I've gotten with everyone here and our upcoming forecast doesn't show the ability to make up the difference before year end."

"Well, the facts are we would be facing that problem one way or the other, since the obsolescence in question was overlooked in our previous forecast. So, we'll just have to bite the bullet on that matter," noted Jim.

"I suppose so, but Mr. Pearl isn't going to like it and if I had my guess he'll go straight to Brooks about the matter; and you know what that means."

"Not exactly, but I'm guessing he'll want me to explain everything."

"More than that," exclaimed Alice.

"In what respect?" asked Jim.

"He'll insist that Harold play a part in explaining it and Harold isn't going to be a happy trooper about that. By the way, have you made him aware of this?"

"No, not yet. I wasn't going to go to him about it until I was sure we had a problem we couldn't overcome. That's just not my style," Jim noted.

"Oh my," replied Alice.

"What's the matter?"

"Well," she started, pausing briefly, "I understand Mr. Pearl is planning to see him today about it."

"Really? Why?"

"I guess he feels since Harold is a peer and you report to him that he should make Harold aware of the situation."

"Thanks for letting me know about that," said Jim. "I guess I better call him before Pearl, if possible."

Alice nodded in agreement and Jim immediately asked Linda in to see if she could arrange a meeting with Harold.

"And Linda? See if you can discreetly find out if he's already met with Justin Pearl today."

Linda returned and said Harold was in a meeting with Pearl at that moment. She had scheduled Jim's meeting with Harold for 3:00 p.m., which was roughly an hour away.

"Great," Jim noted after Linda had left. "Just what we need."

"I'm sorry," noted Alice.

"You have no reason to be sorry, Alice. In hindsight I should have let him know about it before now. I was just hoping we could work this out without a case being made of it. But, unfortunately that didn't happen. I'll just take my medicine, so to speak."

Before Alice could reply, the phone rang. Jim picked it up and answered before remembering that he and Linda had agreed she would take all initial calls. He heard her hang up and thought: *Now I'll have to apologize to her, along with what seems to be everyone else at the moment.*

"Jim, this is Harold. I have Justin Pearl with me and I was wondering if you could break loose and drop by for a few minutes?"

Jim told Harold he was in a meeting with Alice and if it had to do with finances, he wanted to bring her along as well. Harold said to hang on, spoke to Justin, and then replied that would great and to drop by immediately.

"I hope you don't mind," Jim said to Alice as they headed for Howard's office.

"Of course not. In fact, I'm honored that you want me along."

When they arrived, Justin was just leaving having been called to an urgent meeting with Brooks. He and Jim shook hands in a dual hello, farewell fashion. Then, he and Alice politely waited until after Justin departed to take a seat.

"Jim, I was just informed you've missed your forecast for the month by $22,000. That's a big miss."

Jim turned to looked at Alice for a moment, then back at Harold.

"As far as I know we didn't miss our forecast this month, but we have been dealing with the matter. If anything, we underestimated the year-end forecast on our last pass and could have to add something amounting to $22,000 in on the next round. It all has to do with Aklin obsolescence."

"But the facts are you're $22,000 over your forecast and it doesn't matter if it's against the current month or year end," Harold sternly noted.

"I beg to differ. It does matter," challenged Jim and continued before Harold could respond. "If it's a surprise or embarrassment to you, I'm sorry. But

I didn't come to you about it because we found a way to delay some expenditures and take some measures that will allow us to meet this month's forecast. In fact, we may actually beat it somewhat. Now, regarding year end, we haven't as yet submitted our latest forecast and I haven't given up on doing some things to offset this unexpected expenditure. Finally, when all's said and done, this isn't an operating expenditure to begin with. It's an obsolescence write-off created as a result of a customer taking its business elsewhere and on us being required to stock certain levels of inventory, whether it was needed or not."

There was a deafening silence in the room. Harold sat and simply stared at Jim for a long moment, and Jim noticed out of the corner of his eye that Alice was shifting nervously in her chair.

"Alice, if you'll excuse us, Jim and I have something to discuss. We'll get back to the forecast later. Is that all right with you?" he asked.

She agreed, retrieved her briefcase, and left the room, closing the door behind her. Jim was thinking that a battle had been brewing between them for quite some time and it was about to be waged. He decided this was something he wasn't going to back down on, regardless of the consequences. Harold sat with his hands folded in front of him on the gigantic mahogany desk, looking at Jim in a very serious manner. Jim returned the look, determined to wait on Harold to strike the first blow. It didn't come.

"Jim, I have something I need to tell you," said Harold.

Jim nodded an affirmation, but remained silent, not knowing what to expect.

"First, I want you know I agree with you totally. Pearl came running in today screaming as he usually does about how inept everyone else is when it comes to the financial side of the business. As usual, he didn't tell the whole story and I'm glad you straightened things out. But, what I really need to tell you goes beyond all this petty stuff. Jim, I want to let you know that I'll be leaving the operation soon."

Although Carl Armstrong had forewarned Jim, he was unable to respond immediately.

"Harold," he finally got out, "I don't know what to say. But, I'm assuming you're moving on to bigger and better things."

"That's a matter of speculation, I suppose. If you mean to another job immediately, that won't be the case. Brooks and I have agreed I would move on and, frankly, it's something I'm actually looking forward to. What I'm planning on doing is starting my own manufacturing consulting business."

"It sounds interesting. Just what sort of consulting are you considering?"

"It would be broad based and involve almost every facet of manufacturing. What I plan to do is assemble a team of experts in the various functions, ranging from purchasing and accounting to general plant management. I've got some feelers out and the opportunities appear to be good."

"Well, that's great and I'm happy for you. I'm sure it will be a total success. But, again, I'm sorry we're not going to be working together. The truth is, we may not have always agreed on everything, but I've enjoyed the freedom and flexibility you've given me. I'll always appreciate that."

"Not so fast, Jim. There's another reason I'm bringing this up now and I have to ask that you keep what I've told you, as well as what I'm about to say, strictly between us."

"Of course. Certainly," Jim responded.

"Maybe we haven't had the last opportunity to work together," noted Harold, pausing briefly to study Jim's expression in response to the point he had just made. "As I said, I will be putting together some highly qualified talent and I'd like to have you as a member of the team."

Jim was a bit taken aback, but managed to reply. "I'm really honored that you would think of me, especially since you have a number of plant managers who have a great deal more experience."

"I've seen good things in you from the start," Harold responded. "You're a special talent, in a number of ways. But frankly, I sometimes wondered if you had the gumption to stand up to others when you were right. I think you passed that test today and when you did I decided then and there I wanted you on my team. Before we take this too far, I assure you I'm not going to pry an answer out of you now. It would be a big move for you and there would be some risks involved. So, it's something we need to discuss later and not here. You'll need some time to mull it over seriously."

"I fully agree," replied Jim.

"Now, regarding the obsolescence issue, you're right. There's no need for alarm and that's what I'll tell Pearl. If he wants to make a big deal of it and we have to go before Brooks, you can depend on me supporting you a hundred percent on the actions you've taken. Plus, I want you to take advantage of this with Alice and tell her you persuaded me to support you on the matter. I'm sure at this moment she is probably telling everyone you're getting your ass kicked in my office. It's been my experience she's quick to spread gossip. Given that, she'll just as quickly tell everyone how you stood up to me and won the battle."

"I appreciate that," Jim noted.

They finally agreed to meet and discuss a potential consulting role at Harold's home, the following Saturday evening. Harold asked that Jim bring Ginger along for the visit.

That evening, Ginger listened intently as Jim related what had happened in Harold's office and Harold's invitation for them to visit on Saturday to discuss the matter further.

"I have to admit, I was surprised for many reasons. My lingering impression of Harold was he was politically driven. Though there might be some truth to that, given it's easier for him to react the way he did now that he's leaving, I saw a different person today."

"Are you seriously considering his offer? That would be a big step, for the both of us," Ginger remarked.

"Oh, I don't know, but I would have to know more before I say yes. Of course, you would have to feel reasonably comfortable with it. However, consulting does offer some distinct advantages in terms of income, especially if you can get your foot in the door of some the larger firms. Harold said he has some good contacts established, so bottom line, I think we should hear him out and go from there."

"That's fair enough. It never hurts to listen," Ginger agreed.

Harold had been right. The following day, Jim had hardly settled in when Linda stepped into his office and said, "Just thought you should know. The rumor mill is working overtime about your meeting with Harold yesterday."

"Oh? And how's that?" he inquired.

"What I'm hearing is Harold accused you of some double talk and you unloaded on him about how he and Pearl were totally wrong. Harold then excused Alice from the room so you two could go at it without any further embarrassment to her, and you apparently won the argument."

"Well, as usual when it comes to rumors, there's normally both fact and fiction. However, we'll just let it go at that," Jim advised.

The rest of the day was uneventful, with the exception of a special meeting Jim held with the principal parties involved regarding how they would approach cleaning up the Aklin obsolescence and meet the original forecast for the year. Again, he was impressed with Alice's input and suggestions. While there was obviously some truth to her quickness to spread the word on things, Jim wasn't convinced she was a rumormonger. In fact, he and Alice had discussed numerous confidential matters and Jim couldn't recall once when she had broken a confidence. But, he decided he should keep an eye on it.

He kept thinking about Harold's proposal. Down deep, it appealed to him. Though he was still new to the plant management ranks, he had already made up his mind he didn't want to spend his entire career in this all-consuming role. Having been on the job for only seven months, he had already found it extremely taxing because it took a lot of time away from home, friends, and hobbies. Prior to accepting the job, he had enjoyed a regular weekend get-together with his golfing buddies and, occasionally, found the time to go fishing (a passion since youth). Since the move, he had played one round of golf and that was during the off-site staff session, and he hadn't found the time to go fishing at all.

On the other hand, even with all the pressure of the job, he thoroughly enjoyed his work and desired to increase the operation's success, if it was at all within his power. If he chose to leave now, he felt he would be letting everyone down. Plus, he had concrete goals, and such an opportunity didn't come along everyday. Of course, with a baby on the way, he had to remind himself what would serve his family best. Bottom line, he would have to review this carefully and was looking forward to hearing Harold out on the matter.

Jim and Ginger had arrived at Harold's home, which was impressive. Harold's wife, Helen, was a warm and friendly host and spent the first 20 minutes showing them through the place, pointing out the things they took some special pride in and showing them pictures of their children and grandchildren. Harold and Helen had two sons and a daughter, all of whom had finished college and had their own families.

Later in the evening, Harold and Jim settled in the den, next to the warmth of the fireplace and enjoyed a drink while Ginger and Helen were talking in another room.

"Well, Jim," started Harold, "I suppose you've given some thought to what we discussed."

"I'd be lying if I said I wasn't interested. But, as I'm sure you can appreciate, I would need to know more regarding what you precisely have in mind for me and how you plan on making the venture a success."

"It's all very straightforward, I can assure you," replied Harold. "Regardless of whether my departure comes as a result of how it's all unfolding now or some other way, I've been planning to strike out on my own and open a consulting firm. I firmly believe that those I choose to assist me will all be millionaires within three to five years."

"Millionaires?" said Jim, surprised.

"I'm convinced of it."

"There's that kind of money in general consulting?"

"Doing it the way I'm planning, yes," Harold noted.

"And how's that?" Jim pressed.

"Well, I wouldn't want to give all my trade secrets until I know for certain someone's fully on board. You can probably appreciate that."

Jim nodded his agreement and Harold continued.

"It's going to work something like this: I'll do all the marketing and contracting of the work myself. That will be my principal role. Depending on the need expressed by the customer, I'll assign one or more of the partners to the account and it would be their responsibility to conduct the on-site work. However, rather than the customer paying a flat consulting fee, our initial service will be free. Our financial reward would come as a result of what we saved the company and that would work something like this. I would get an upfront agreement for 50 percent of whatever we saved them over the first two years. For example, savings in areas like scrap, obsolescence, and rework, as well as manpower reductions from productivity gains. The key is to establish an agreed-upon base from which to measure progress, but I don't believe I'm going to have any problems doing that. It's basically a straightforward affair. The principal selling feature is that our customers aren't out anything initially, other than travel and living expenses for our consultants, and when it's all wrapped up we get a 50/50 split on the savings. Bottom line, if we do the kind of job I believe we can, we would be talking about some lucrative income for everyone."

Harold paused for a moment to let what he had just said sink in and then continued.

"Now, as founder and owner of the consulting firm, I would take 50 percent of the income, with 40 percent going to the partner or partners performing the on-site work. The remaining 10 percent would go into a bonus pool that would be equally distributed to all the partners annually."

"It certainly sounds interesting and the approach is unique to say the least," Jim noted. "Who have you spoken with other than me and what's been the response? I ask this because the results are only going to be as good as the team you put together. Right?"

"You're right. So, to answer that, let me say that I'm working on it and would prefer not to give any names at this point. Due to the nature of this I wouldn't want to get any unnecessary rumors started. But I can assure you they would be people I highly respect and whom I've worked with, both here and elsewhere."

Almost as if lightning had struck, Jim felt some doubt creeping in. If Harold could trust him to discuss a potential offer and the details of his approach, then why wouldn't he trust him to keep names in confidence? And what was his comment regarding unnecessary rumors supposed to mean? It didn't make a lot of sense.

"Understanding who my associates are going to be would be important and it would, of course, be difficult for me to make a commitment without knowing that. I am sure you can appreciate where I'm coming from because we're talking about a big step here," Jim noted.

"Of course," Harold replied. "And you would know that soon enough should you decide to make a tentative commitment. But I want to let you know what a solid commitment would mean, in terms of stock in the partnership."

Didn't he hear what I just said? Jim thought. *Now he's talking about a price for a partnership.* However, Jim decided it would be best to just listen at this point and keep his mouth shut, but down deep he didn't like the way things were proceeding.

"Please go on," he urged.

"I'm sure you can understand. I need people who are committed to this venture and, since the financial rewards will potentially be high, it would only be appropriate for those joining to show a solid commitment by putting in some stock. Bottom line, a partnership is going to cost $30,000, which is actually a small price to pay considering the potential rewards. But I want to give those who are helping me start this venture a lasting benefit. What I mean by that is should anyone decide to leave the firm, at some point, they can sell their partnership for any price they chose."

"I see, Jim replied, doubt now seriously beginning to set in. "But, considering someone decided to do that, would there be any restrictions on whom they could sell their partnership to?"

"None whatsoever," replied Harold.

"But wouldn't that pose a problem? I mean, shouldn't you have some established guidelines regarding the qualifications of a person before allowing them to buy in."

Harold chuckled. "If they have a hundred grand or whatever it takes at that point, I'd say they're qualified."

His last comment literally stunned Jim and in essence pulled a veil of suspicion over the entire matter. It seemed that what Harold was proposing was little more than a get-rich-quick scheme, which was heavily skewed toward Harold. Jim decided then it wasn't something he wanted to rush.

They finally decided to call it a night and with an appropriate farewell, Jim and Ginger departed. The only thing solid Jim left with Harold regarding what they discussed was that he would give it some serious thought, talk it over with Ginger, and get back to him. But the truth was he had already made up his mind.

"Well, how did it go?" asked Ginger as they were driving home.

"Ginger, you wouldn't have believed it if you'd heard it. It was humorous, if anything. Harold has cooked up this cockeyed scheme and I will be surprised if he gets any takers."

"Oh?" she asked.

"Guess what he's offering me a partnership for?" he replied.

"What?"

"Thirty thousand dollars," he noted.

"What?"

"That's right, honey, thirty thousand. If I give him that, he'll allow me to be a part of the team and bask in the rewards. However, he refuses to tell me the names of the others he's thinking about offering a partnership and he's placing absolutely no restrictions on whom they could sell their partnership to. That means anyone can bail out any time and anyone who has the money and the inclination, regardless of background and qualifications, could become a full-fledged partner. Frankly, it's about as absurd as anything I've ever heard. Bottom line, I see it as a means for Harold—who isn't stupid by the way—to bring four or five suckers in and end up putting a few hundred grand in his bank account for leaner times."

Ginger listened politely as Jim continued.

"The thing that makes the entire proposal ridiculous, in my judgment, is that no consideration is being given to maintaining the kind of talent that would be needed to make it long-term success. Considering he did recruit some decent people on the front end, which at this point I'd have to doubt, there's nothing to keep him or any of the others from taking a gain and bailing out at any time. And once the appropriate talent is lost, the partnership would be totally worthless."

"Did you mention that to him?" asked Ginger.

"No. It became clear to me where the venture was headed, so I decided to listen and keep my mouth shut."

"Is that fair?" she challenged.

"What do you mean?" he asked.

"Have you considered that it could be nothing more than honest but poor judgment on his part? This is something new for him, as well as you. As a matter of fact, Helen and I discussed that and she mentioned how he had been struggling with how to go about organizing things. I like her, by the way. Anyway, it could be that he just needs someone to point out what you've mentioned and offer some constructive suggestions. After all, Jim, you said yourself that this could be an opportunity of a lifetime. However, I'd be the first to admit that thirty thousand dollars is a lot of money. That would be a big commitment for us."

"I don't know," he replied. "Frankly, the real question could be why I would even consider discussing something like this with him in the first place. I admit I haven't trusted him and I've got nothing to base that feeling on other than his reaction to certain things, but I believe he's politically driven. But that impression could be wrong."

"Of course, it could be wrong. You have to admit you're basically a suspicious person. Maybe that's what makes you the ideal plant manager. So, I'd say give him a benefit of the doubt."

"Oh really? Suspicious?" he joked, glancing briefly at her and smiling broadly before turning his attention back to the road.

"Yes. Suspicious," she responded, scooting closer and giving him a quick peck on the cheek.

Ginger never failed to amaze him. She had a way of viewing things with a good deal of commonsense. He was fortunate to have her and although he didn't always like her ability to pinpoint his weaknesses and her quickness to point them out, he knew anything she offered in criticism came from a deep concern for him. After brewing over her input, he responded.

"You know. You're right. Since he's taken enough interest to approach me about the matter, I should stay positive about his motives, up until he proves otherwise."

"Exactly," noted Ginger, and with that they dropped the subject.

The following day, Jim, Harold, Phil, and Norm Bowen settled down to a meeting in Harold's office. Bowen was Denning's chief corporate negotiator. He spent all his time on the road, working with the various operations on contract negotiations and, when questions or debates arose, the interpretation of contract language. He was the company's guru when it came to anything associated with contracts and labor issues.

Bowen's official role was to act as chief liaison between the company and the union's senior advisors and elected leaders. Unofficially, however, he called the shots on the company's position for pay and benefits. He also determined where the company was willing to withstand a strike, if necessary, in order to maintain some reasonable continuity in its union contracts. Essentially, he carried a lot of power and influence.

"We're here today gentlemen," he began, "to sort out what's going to be important, from your viewpoint, for us to take to the table in the upcoming negotiations. Actually, we're a little late in starting this dialogue, since we're less than six months away from our initial meeting with the union. Here's your copy of a confidential document that outlines Denning's basic position on any future contracts. Make sure you keep this in a safe place and out of the hands of those who don't have an absolute need to know. I'm sure you can understand why."

Jim, Harold, and Phil nodded as they studied the document.

"As you can see, this comes from Brooks and it's just about as straightforward as it can get. It says there are three key objectives that every manufacturing operation is going to pursue, regardless of their particular union," he noted.

Bowen reviewed the document in detail. It stated a total dollar value, in terms of pay increases and other benefits, not to be exceeded. This was roughly 15 percent less than the going average of all the existing Denning contracts. Second, specific labor classifications would be lowered from hundreds, in some cases, to no more than 25 in any operation, preferably even lower. Finally, under no conditions would elected union officials be exempt from holding down a job on the factory floor in order to work full time on union business. Bowen went on to note that most of the plants in the past had allowed at least the president of the union to spend full time on union

business and Brooks was adamant about ending it. When Phil Tanner interrupted to express concern for what seemed to be a petty, if not highly ques-
tionable objective, Bowen responded.

"Brooks and I got a lot of input on this from Patricia Owens, who as you
know he highly respects, especially since she grew up with Barring Corporation.
Brooks definitely thinks Barring is at the head of the heap when it comes to
top-notch companies and, of course, he picked Patricia's brain on this matter."

In looking around, Jim could see the mere mention of Owens didn't set well
with the crowd. In fact, Harold didn't try to hide the precipitous frown that
crept across his expression as Bowen announced Owens' participation in the
matter. He finally spoke up.

"Well, I have to tell you, with these objectives, I hope they discussed a
strike plan," he said, clearly sarcastic.

Bowen shrugged his shoulders and ignored the remark. Jim could tell
Bowen knew Harold was on his way out and had chosen not to say anything
that could stir up an unnecessary confrontation. By handling it the way he
did, he sent a strong message that Harold's comments weren't all that important, and the only reason Harold was there was as a courtesy, considering his
current title and position.

"Now," Bowen continued, "what we need to discuss are the issues you
want to address at the table and the things you feel we should accomplish. I
have to tell you, however, that with what we're going after this time, any
other significant changes are going to be difficult to achieve."

"There is one thing I was hoping we could go after," Jim noted. "And, this
was to pursue a four by ten work schedule. Harold and I have discussed this. It
was something we negotiated at Crafton, where I previously worked, and it
went well. I assume you know what I'm referring to, Norm. For the others
here, I'm speaking about a four-day work week, with a ten-hour per day work
schedule for everyone."

"I'm familiar with it," replied Bowen, "but I've got to tell you, we don't have
a snowball's chance in hell of pulling something like that off, especially considering the key objectives Brooks is after."

Jim decided to press on. "I'd have to say that some of what you're telling
us generates serious questions. First, lowering classifications is admirable, but
most of them are created at the insistence of the company, not the union.
Once they're in place, taking a majority of them back can stir up big problems.
I know because I've been involved with something similar before. To lower
classifications without clearly understanding how work gets done, and by
whom, will cause real misery. I'm not ignoring we should lower classifications,
but we have so little time before negotiations to get our house in order for
something like this. It could end up creating some serious problems for manu-

facturing. I'd say it's something we should consider down the road when all the plant managers have had more time to work on it. Regarding forcing elected union officials to hold down full-time production positions, I'm not against it in principle, but you're looking at peanuts in financial terms verses all the controversy, especially with us going after some reductions in anticipated pay and benefit increases. So, the question becomes, what purpose would it serve?"

Bowen thought for a moment before replying. "Well, Jim, I've got to tell you, Brooks isn't going to delay or drop any of it."

"But if that's the case, it doesn't leave a lot of room for anything else," Phil noted.

"Then our meeting today should be short and sweet, wouldn't you agree?" Norm responded, looking around at everyone and smiling to punctuate his remark.

"It certainly looks that way," Harold interjected, disappointment clearly registering in his voice.

It became evident to everyone that the purpose of the meeting really wasn't to discuss what manufacturing felt should to be pursued at the table, but rather to understand what was going to be dictated. When nothing further was immediately offered by anyone, Norm closed the discussion. "In that case, I guess we can call a halt to this and get back together in a week or so to discuss logistics," said Norm.

With that the meeting was wrapped up and Norm quickly departed. Jim, Harold, and Phil sat for some time, continuing to study the details of the document. Finally Harold stood and pitched the document toward his desk. It drifted in the air momentarily before settling quietly on the slick glass top covering the desk and then slowly slid off onto the floor. Harold didn't bother to pick it up.

"I'm far too upset to continue any kind of constructive dialogue now, so we'll get back together on this later. I'm seriously offended that a directive was put together regarding how we're expected to run our end of the business with zero input from any of us. However, I need to share something with you that could shed a little light on the whole affair," he said.

Harold opened the desk's center drawer and retrieved another piece of paper. He handed it to Jim, asked that he read it first and then allow Phil to look it over. It was a draft copy of an announcement from Phillip Brooks:

Harold J. Jenkins, VP and General Manager of Manufacturing for Denning Corporation, has decided to pursue a career interest in business consulting. As Harold has completely assured me, this is something he has long desired to make a reality. Therefore, it is with profound respect for his past achievements and sincere regret that he will be leaving us, that I make this announcement.

In conjunction with this and in order to enhance our overall organizational structure, Patricia M. Owens is promoted to the position of VP of Quality and Manufacturing. In this newly established position, Patricia will serve in a duel role as both Denning's quality assurance director and manager of manufacturing operations. This will provide an emphasis on coupling total quality assurance with enhanced manufacturing methods and procedures. In turn, it should improve product quality and enhance both delivery performance and customer satisfaction....

Jim finished reading the draft, passed it on to Phil and then sat quietly as he gave him time to look it over. After a few moments, Phil handed the announcement back to Harold.

"This is going to be posted tomorrow, so I wanted to forewarn you and discuss how we should distribute this information to all the different operations. Phil, I would begin by telling you that I have already made Jim aware I was going to be leaving. However, what's new to Jim in all likelihood is that Patricia Owens is going to be taking over, although that may come as no big surprise. I learned this myself only yesterday."

Phil extended his hand to Harold. "I assume congratulations are in order," he noted, moving on to something more positive.

"Thanks. It's something I've been thinking about for quite some time," Harold replied, giving Phil a warm handshake before proceeding.

Harold went from there to discuss how and when the announcement would be distributed and posted throughout the various plants. He noted he still had to get with the other operations and make everyone aware. Consequently, he asked that they not divulge the information until he had the chance to do so. They all agreed and Harold then excused Phil, requesting that Jim stay for a few moments longer.

"Well, what's you're reaction to all this?" asked Harold.

"Disappointment," Jim started. "To begin with, I'm not a big fan of Owens. But the feeling is probably mutual. She and I just haven't hit it off well. She gives me the impression she's blinded to anything other than her own personal aspirations. We'll just have to see how that goes. Regarding the contract, I was really surprised and agree with your displeasure over the way the entire affair has been handled."

"Have you given any thought to what we discussed the other evening," Harold asked, quickly moving to what was more important for him.

"Certainly. A great deal. But not without a few reservations to be honest," Jim responded.

"I expected that. It's a big step to say the least," replied Harold.

"Since you brought it up, Harold, I'd like to pursue some things further with you about the matter, but this isn't the time or place."

"I agree. How about us getting together for a drink after work? Markers has a nice lounge near here."

"Sure. I know where it is. Say about 6 P.M.?" asked Jim.

"Perfect," replied Harold.

Later that evening, Jim and Ginger sat in their den discussing the day's events. He told her about the pending announcement and his subsequent meeting that afternoon with Harold at Markers.

"I took your advice and expressed my concerns to him, as well as tossing out some suggestions. I really don't know how to define his reaction, other than to tell you he was less than responsive," Jim noted.

"In what way?" Ginger quizzed.

"He listened politely enough, but he didn't make any strong commitments one way or the other. However, his basic position was my chief concern was ill-founded—you know, about people potentially bailing out for the money and leaving the firm with less than adequate talent. He said that wouldn't happen. I'm still not convinced. When I asked him again about whom he was considering, he made it clear he wasn't going to divulge that to anyone who hadn't already bought in. Of course, I let him know I'd have a hard time saying yes without knowing who my associates were going to be. I really can't understand why the need for this wouldn't be totally obvious to him. He did say, however, that he was limiting the total number of partners to four, including himself and that he had some talented people expressing interest."

"Interesting," noted Ginger. "So, where does it go from here?"

"I'm not sure. We just left it that I'd give it more thought and get back to him."

"Hm. I'd say you should give him a little time to mull it over. And let him come back to you rather than the other way around. If he's interested in having you as a partner, he'll do the right thing. But being unwilling to give you even a hint regarding the others he's considering is strange, to say the least."

"I'm thinking about telling him to rule me out. I got far too many reservations at this point and if I harbor some doubts now, it's probably not going to get a lot better with time. With our baby on the way, I have to think a lot harder before taking a career step that could backfire before it's all over," said Jim.

"You're right, but what harm is there in waiting him out a little longer? You haven't made any solid commitments, so it's no skin off your back to wait and see what develops. Wouldn't you agree?" Ginger advised.

"I suppose so, but I wouldn't want him to get the wrong impression about how I'm leaning on the matter."

"Well, I think you've made that clear to him," she said. She suggested he sleep on it and, with that, they went to bed. Even in his sleep Jim couldn't get work off his mind and dreamed about Patricia Owens. She had asked him to drop by what had previously been Harold's office and when he entered she was seated behind the desk, doing paperwork.

She had changed the office's decor drastically. On each of the four walls were some exceptionally large murals, each containing a DQS2000+ theme or message. Patricia was dressed in an imposing dark business suit, with her hair pulled back and pinned tightly to her crown unlike how she normally wore it.

"Have a seat," she instructed as she continued to shuffle through the papers, waving a hand toward the two chairs in front of the desk.

Jim sat down and studied the unusual decor as she continued with her task. She pushed the papers aside, leaned back in the chair and folded her hands neatly in front of her. After a moment, she turned slightly to her right and bent to retrieve something from the floor before straightening again to look at him sternly. She slowly rose from her chair and stepped to one side of the desk. Jim was stunned with what he saw. In her hand, she held a long, coiled whip.

"Stand up," she ordered.

Jim hesitated slightly, but then stood and faced her.

"Look at me," she commanded.

"What's going on?" Jim asked. He was perplexed and starting to feel uncomfortable with her general demeanor and unusual conduct.

"Shut up! And don't make me give you an example of what I'll do if you don't follow my orders," she demanded.

Jim's anger rose, and he almost told her to shove it and walked out, but decided against it. He surprised himself by nodding.

"We're going to put some things in place to see that manufacturing is run my way. Do I make myself clear? Don't just stand there, speak up."

"Yes," he reluctantly responded.

"You'll forget about all the petty junk you think is important and focus instead on what I want to accomplish. Understand?

"Yes."

"Yes, what?"

"Yes. Master?"

"No," she screamed, stepping back and lifting the whip in the air with her right arm before allowing it to uncurl behind her. "It's Mistress, you idiot!"

Then, in what seemed to be slow motion, the tip of the whip started to descend toward him. He tried to leap to one side but his legs wouldn't respond. Just before it struck the arms he had extended above his head, he was in an excruciating free fall, inside what seemed to be a bottomless pit. He felt completely helpless and fear started to grip him as he fell faster and faster.

Jim awoke startled and confused and found himself sitting up in bed. It took a moment for him to realize it had been a dream, which was something he hadn't experienced since he was a child. He suddenly felt relieved but foolish about how absurd it had been.

After a while, he settled back in bed in an effort not to disturb Ginger, who was still fast asleep. As he rested his head on the pillow, he realized it was damp with perspiration. He proceeded to turn over and rolled onto his side facing the dual windows of the bedroom, where a dim ray of light drifted through. He finally fell into a relatively peaceful slumber. The next morning he shared his experience with Ginger during a light breakfast before heading to the office.

"I haven't had a nightmare since I was kid, and it was a strange one," he said.

She laughed warmly as she stood and proceeded to the kitchen to reheat her coffee. After she returned and took a seat, she added, "She had a whip, huh? How interesting."

"Now, Ginger," he responded with a smile.

"Well, you know what they say about dreams," she joked. "Maybe you have this kinky side you've been hiding from me all these years."

"Enough, please," Jim remarked, proceeding to give her a peck on the cheek before departing.

"On second thought, it's probably just a macho thing. You know, having to report to a woman and all," she continued.

"Stop it." He was suddenly wishing he hadn't said anything. Ginger could be a real joker sometimes, and he was sure he hadn't heard the last about this.

As he started to close the door behind him, he heard her say, "But if you see her headed your way carrying a set of handcuffs, run like hell."

"I'm definitely out of here," he chuckled.

Jim was holding his biweekly meeting with Mark Hudson, the president of the local union and two of Mark's elected officials he had decided to bring along. The meetings over the past several months had gone well, in general, and had proven to be a catalyst for resolving some issues that would have otherwise been a problem. Jim was glad he had started the dialogue.

"We have our preliminary contract demands ready for the company to look at when you're ready," noted Mark.

"I'd prefer to bring Phil in on that, but I'll tell him you're ready to discuss it."

"We assume you'll be letting us know who's going to be the chief negotiator for the company. We'll, of course, be bringing in Conrad Mason as our chief consultant. He's our man when it comes to these matters."

"I don't believe I've met him. I assume he's from your international headquarters. What's his title?" Jim asked.

"He is from headquarters and you haven't met him, as far as I know. He's a vice chairman and works exclusively on contract negotiation."

"Good. I'll be happy to meet him. Do you plan a get-together between the parties prior to kicking things off?"

"That's been a long-standing practice. Usually, it's a dinner affair so everyone gets the chance to meet and know everyone else before official negotiations begin," said Mark.

"And who starts the ball rolling on this?" asked Jim.

"That's always been up to Norm Bowen. Hasn't he spoken with you about it?"

"We've had some conversation, but nothing, as yet, about this dinner."

"I'm surprised," Mark said. "You know, we're less than six months away and usually we've already had three or four preliminary get-togethers by this time. I've let Phil know my concern about this, so I'm assuming he hasn't bothered to pass that on to you either."

"Honestly, I don't recall him mentioning it to me. He could have, but it isn't like me to forget something like that. However, I'll check it out and let you know where we go from here. Hopefully, this hasn't created any stumbling blocks for us."

"Fair enough," noted Mark.

Jim wanted to take advantage of having them together to pass on the announcement about Harold, but it had been agreed to be done in a special session including Phil, just before the announcement was posted. It was scheduled for 3 P.M., which was approximately four hours away. They wrapped things up, and afterward, Jim dropped by Phil Tanner's office to relate his conversation with Mark. He then waited for Phil's response.

"Well, yes, Mark has approached me a couple of times wanting to know when we'd be ready to take their initial demands and when we planned to get together as we normally do for dinner. But the facts are, Norm hasn't been very responsive when I've approached him about the matter."

"In what respect?" asked Jim.

"In every respect. The standard response I'm getting is he's working on it and when he has things sorted out with senior management, which I assume means Brooks and probably Patricia Owens he'll get back to me."

"When was the last time you two discussed this, Phil? The reason I'm asking is to determine if it was before or after our meeting with him."

"Oh, it was after. In fact, just yesterday," Phil replied.

"I see," noted Jim, pausing. "Here's what we'll do. I'll call Norm myself and ask for some time, so we can get some answers. I'll delay doing this until after the announcement on Harold and Patricia since I'd also like to speak with her about it."

"Sounds right to me," replied Phil.

Before Phil departed, Jim asked to be told if any further conversations developed with either Mark or Norm about the issue, and Phil assured him he would.

Later that afternoon, the announcement about Harold leaving and Patricia Owens assuming command was posted, and within an hour the entire place was buzzing about it. The rumor mill cranked into gear, and before the day was over, Alice let Jim know a story was going around that he was going to be replaced soon since most everyone knew Patricia and Jim didn't get along.

"It never fails to amaze me how rumors start and grow around here," noted Jim. He thanked her for telling him. Although he wasn't always receptive to Alice's rather quick propensity to pass on the latest gossip, she seemed to feel it was her duty to do so. And, there was a value in knowing what was going around in order to avoid surprises at the wrong moment. He had to be mindful of any response he made since he hadn't decided how much he could trust her to keep things to herself. Better safe than sorry, he decided. Just before wrapping things up for the day, he got a call from Patricia Owens' secretary, asking if he could meet with Patricia at 8 A.M. the following day. He agreed.

"Any idea what it's about? The reason I'm asking is to understand if I should come prepared to discuss something specific," he asked.

"She didn't say," her secretary replied as if she were offended he would ask.

"All right. I'll be there."

The next day, as Jim walked into Patricia's office, he recalled his nightmare and proceeded to chuckle about it as he settled into the chair at her conference table.

"Well, you're in a good mood."

"I suppose so," he replied. "And how are things with you?"

"Just fine, thank you," she remarked before turning to business.

"Jim, I've asked you to drop by so we can discuss a few things."

"Fine. Before we start, let me congratulate you on your promotion. It's a big job you're undertaking," he diplomatically interjected, extending his hand to her.

She thanked him again and returned his offer for a handshake.

"You can depend on me doing all I can to help," he went on.

"I'm sure you will. Now, getting back to why I've asked you here, I've decided on a couple of things that directly affect your operation, and I want to make sure you're up to speed."

Jim remained silent during a short pause as she searched for some notes, which she apparently wanted to have at her disposal.

"Oh, here it is," she said, before leaning back in her chair.

"I've decided to speak with personnel at each of the plants over the next few weeks, starting with your operation. I'm doing that because you're here at headquarters and you can understand the importance of putting things in that order. It's going to be a sort of a state of the business address and along with that, I intend to outline our mission for the future."

"I see," replied Jim.

"Regarding the condition of the business, I'm going to hit hard about how poor quality at the factory level being the principal factor in us losing Aklin's business. Brooks, as you know, is quite upset about it and he's personally challenged me to see that something like this doesn't happen again."

"I understand and I believe we've made some good strides in that direction," said Jim.

"Oh? That's a matter of opinion, I suppose. Your dual inspection process has admittedly put a patch on the problem, but we have to go much further than that," Patricia responded.

"I wasn't referring to the inspection process. I was referring to the initiatives we've started that are designed to get to the root of the problem," replied Jim.

He pointed out he had recently started a full-blown continuous improvement process they were calling the Denning Production System, and that training and floor activities were underway. He further noted some of the initial progress was being made, in terms of process improvements, freeing up floor space and cutting scrap and rework.

"You've put something in place you're calling the Denning Production System?" asked Patricia, obviously disturbed about it.

"Yes," Jim replied. "Is that a problem?"

"I wish you hadn't done that and, frankly, I'm surprised I wasn't made aware of it before you kicked it off. In all fairness, let me explain where I'm coming from," she said.

Jim felt like asking how he was supposed to know she would be taking over when he, Harold, and others decided to move forward with the initiative some three months ago, but he decided to hold off a challenge for the time being.

She restated the importance of DQS2000+ for what she had told Brooks she would accomplish and then surprised Jim by saying she would tell everyone it was to become the *production system* throughout Denning.

"The production system?" Jim replied.

"That's correct," she responded.

"But how can that be? DQS is a quality system and though it can contribute to the way the business operates, it doesn't address other important factors to be pursued in order to make us a world-class operation. At least not in my judgment," he added. But he could immediately tell Patricia was struggling to maintain her composure. He thought *Here comes her my way or the highway speech.*

She began calmly. "We don't disagree in principle. The objective is making this a world-class operation. It's only that I see it just the opposite of you. What you're doing is a good support mechanism for DQS2000+. But, we can't have people confused about our chief priority by calling what you're doing the Denning Production System. That would lead them to believe otherwise."

"I assume when you speak about chief priority that you're referring to DQS2000+?"

"Exactly."

"Well, I feel just as strongly about what the impact would be if we suddenly decided to change the name of the process we've put in place, which, by the way, we have put some genuine time and effort into getting on the right track," he interjected.

"I guess you'll just have to get over that," she replied in a matter-of-fact tone.

Jim leaned back in his chair somewhat, squared his shoulders, and let a breath of air escape in an unconscious display of some obvious frustration.

"So. What are you telling me?"

"I'm telling you that you'll have to correct the situation."

"I don't see anything to correct," he challenged, almost as if baiting her for a battle. And though he knew what it could potentially lead to, he simply couldn't help himself.

Patricia studied him for a moment, her eyes shifting swiftly from his face to the notes she held in hand. She struggled with her composure. After a long moment, she laid the notes on the table and then looked directly into his eyes.

"Jim, I want to appreciate your viewpoint and I want us to be able to work together in an atmosphere where freedom of thoughts and ideas can be freely exchanged. However, there are going to be a few things that simply aren't debatable and I want you to understand that DQS2000+ is one of them. Believe me, I don't want to damage what you've started and I'm glad to hear about it. It's needed, I'm sure. But you're going to have to work with me on this," she pointed out.

He was far from convinced about her sincerity.

"Now, let's see if we can't go at this from another direction. First, we should think about how you would approach the troops on changing the name of the process," she noted.

He wasn't optimistic about what the outcome was going to be.

"What if you got them together and made it your idea?" she asked.

He felt a little better. At least she was considering a constructive change.

"You could even say that the name Denning Production System simply doesn't cover the depth of the mission and that you want to structure it toward a higher meaning," she suggested.

Not a bad idea, he thought.

"Or, you could wait until I make my announcement and tell them that you want to expand the active implementation of DQS2000+ into your initiative and call all that the Denning Production System. That would serve two purposes. It wouldn't do anything to shed a poor light on what you're already doing and would make DQS2000+ a major part of your entire process."

He was more impressed. She was displaying the ability to think things through in an intelligent manner, while keeping a strong focus on her objectives and, at the same time giving appropriate consideration to the feelings of others. He liked what he heard.

"I believe I like the latter suggestion, Patricia," he noted.

"Good. I do, too." She smiled.

On his way home that afternoon, Jim felt better about Patricia Owens. Even though he was the first to admit he was sometimes too quick to form first impressions, he had always been slow to change them. Something special had to happen to change his mind, and it did.

They had spent almost an hour together before he departed. She let him know about the goals and aspirations she had in her new role, yet seemed just as interested in hearing his. She appeared anxious to learn about his background and his family and expressed a genuine interest in what he had to say. Toward the end, she admitted things had gotten off to a rocky start between them, but placed the blame on herself for sometimes letting her drive and ambition get in the way of an appropriate consideration of others. She confessed she understood her weakness and was working on it, and Jim let her know he saw evidence of that.

Later, when he related what had happened to Ginger, she said she was glad to hear it turned out that way. Then true to form, she said, "So, she didn't bring out the whip?"

Jim put on his worst face and then crouched like a bear about to pounce on a victim. She knew the signal and turned to scramble off quickly toward the next room, giving a playful squeal as she did so. He lumbered behind and caught her. Then wrapping his arms around her and growling like an enraged animal, he picked her up and gently deposited her on the couch. Ginger was laughing, so it took her a moment to get out what she wanted to say.

"Careful. The baby," she warned. He was certain she wasn't saying that out of some genuine concern, but rather because she knew what was coming. They had played this little game before.

He flopped down on the couch next to her and began to tickle her with both hands. Within seconds she was squirming and giggling like a six-year-old. After a moment, the two of them settled down somewhat and he leaned forward to give her quick hug before standing and declaring in his deepest caveman voice. "Baby fine. Baby strong, like daddy."

Ginger buried her face in both hands and then peeked at him through a small opening she made by spreading her fingers. She was smiling warmly. He was suddenly thinking, as he held her hand, that he would be lost without her. She was indeed his one true friend and comrade.

AVOID THE TENDENCY TO PREJUDGE MOTIVES

Everyone would agree that getting satisfactory results requires the ability to work effectively with others; and there is a world of advice on how to go about this. What I'm referring to, however, is the single most important factor in getting things done through others as well as building lasting relationships.

I have discovered the key is to keep a clear and open mind as it pertains to the motives of others. My advice on this particular subject comes from the school of hard knocks. Experience has taught me that managing through others isn't the ability to impress and/or motivate. It's a matter of achieving and maintaining a mutual level of trust and respect.

Consider what your success rate has been when you had mutual respect and trust. Assuming whatever you jointly decide to achieve is realistic and you have no doubt as to each other's motives, your success rate is generally going to be high. However, even when failure occurs, there isn't the tendency to place blame. Therefore, walls of uncertainty and doubt aren't constructed that could hinder the ability to work together again in the future.

Now consider the same scenario where a lack of mutual respect and trust exists. The success rate in achieving an agreed-upon goal would be low. Assuming you are not a person who simply bases respect and trust on how others look, speak, or act, then the reason for this has to be some question as to the other person's motives.

However, being human, we have a tendency to prejudge the motives of others. This is true as it applies to those we don't know well. For instance, if a stranger approaches you on the street and asks for a handout, you tend to question what the person wants. In other words, what's the motive? Is he or she in desperate need or is it a scam? Has he or she approached you because of unfortunate circumstances or to set you up for something more sinister?

When you are new on the job, most of your peers and associates are essentially strangers and the only thing that separates them from the person on the street is that you are both employees of the same company. Therefore, there is a certain level of comfort not found on the street. However, this doesn't stop us from questioning the motives of peers and associates. In fact, we sometimes raise this to even higher levels.

So, how do we avoid prejudging the motives of others? Start by reminding yourself to give everyone's motives a benefit of a doubt, without exception. This isn't easy, so you have to work at it daily. However, when I recognized and applied this approach, I was amazed how much more I achieved with and through others. I only wish I had started this much earlier in my career because it would have saved me a lot of unnecessary grief. But, as they say, with time and experience comes knowledge and wisdom.

Seven months later, Jim and Ginger were the proud parents of a bouncing baby boy they named Nathan. Patricia Owens had assumed command and after much deliberation, Jim had decided to decline Harold's offer. He was now glad he did.

Shortly after Harold left, Bob Frisman turned in his resignation to join Harold's firm and while Jim felt Bob had a lot of potential, he wasn't the kind of partner Jim saw as ideal for such a venture. He was too inexperienced and unproven, in Jim's judgment. It only reinforced Jim's opinion that he, too, was far less than an old pro when it came to plant management expertise. Along with Bob, Jim had learned that one other relatively young middle manager from another Denning operation had also resigned to join Harold. Apparently, Harold either didn't have a lot of takers who fit into the highly experienced mold or he was consciously recruiting some extremely young talent. Either way, Jim was glad he had declined.

Patricia had proven to be better to work with than Jim had imagined. Though she had irritating quirks and struggled to control a quick temper, she gave all her plant managers a good deal of freedom to operate. She had shown the ability to be flexible. Jim thought she was merely a pawn to Brooks, but he found this was not the case. On several occasions, he and other plant managers watched her stand tall and challenge Brooks' position. In turn, he was equally quick to openly differ with some of her thoughts and opinions. Much more so than what they had seen in his dealings with Harold. Therefore, it was apparent they had a working relationship where the level of mutual respect was high enough that they didn't feel it was necessary to mesh words. All in all, it had served to set up an atmosphere that was much more conducive to getting concerns out in the open and keeping things on course.

Jim had been forced to make a couple of tough calls. The first was deciding he had to move Joe Thompson out of his current position. He was old school when it came to manufacturing and just couldn't buy into the *Lean Manufacturing* practices that Jim was pushing. Patricia Owens wasn't fond of Joe to begin with and her position had been to let him go. However, Jim was determined to avoid that, if possible. Regardless of what one might think of Joe, Jim thought he had a wealth of experience and could be a solid contributor in the right role. But that role definitely wasn't production manager. He decided to move him into Bob Frisman's old job, with the understanding that he expected Joe to show more support and enthusiasm about the process. Joe hadn't been happy about it initially, but he was smart enough to know that it could have been worse. Overall, he was now performing decently in his new position.

As a result, Jim had Phil recruit and hire a new production manager. A man named Rex Anderson. Rex had been in manufacturing for almost ten years and had served a short stint as a plant manager over a small facility. He had good credentials, with one of the more desirable being his background in leading the implementation an aggressive world-class manufacturing approach, at his previous company. In addition, he dealt with employees and customers well. Rex had now been with the operation for three months and Jim was most impressed with his results. In addition, Rex had even been able to win Joe Thompson over and influence him to make some needed changes in P&S.

The contract with Aklin had expired and the resulting layoff had made the headlines in the local paper. The town of Bristol, where Denning's headquarters had resided for over 20 years, was located in a small mid-western community and Denning was by far the largest employer around. A local television crew arrived at the plant the day of the layoff and Jim ended up giving an unexpected interview, as he was caught leaving the plant. Overall, he had handled himself well, which was evidenced by the fact that both Brooks and Owens let him know they were very proud of the way he conducted himself. Jim's response to the press was Denning was in a competitive business and while the loss of any customer was unfortunate, efforts were underway to gain additional business and get those on layoff back to work, as soon as possible. He continued by noting that most of those affected were temporary employees. The following day, it appeared the newspaper intentionally stirred controversy over Jim's use of *temporary*, and the next day a rather lengthy article followed where a disgruntled temporary employee was interviewed and ended being critical of the company. He bitterly questioned why he was hired only to be laid off a short time later. Of course, he didn't mention that every temporary was informed at the offset it was just that—a temporary position. Shortly thereafter, things cooled down and nothing more was mentioned by the news media.

One of the bigger issues turned out to be the labor contract. The union and the company had failed to reach an agreement and to bring it to a vote by the membership. As a result, they had agreed to a 60-day extension of the current contract, in an effort to avoid a work stoppage and provide some time to resolve their differences. The union had rejected the company's financial package, but everyone close to the matter knew this wasn't their primary point of contention.

The real problem was the company's demand that all union officials be required to hold down full-time production jobs. Union officials were, in fact, irate about it and took the position it would make it impossible for them to do an adequate job of representing the rights of the members. Further, requiring union officials to get clearance before leaving their job to investigate grievances wasn't practical, it was only a ploy by the company to weaken the union's ability to represent its members' welfare satisfactorily. Even though Jim would never admit it publicly, he agreed with their position in principle. He felt a good deal of unwarranted emphasis was being placed by the company on a matter that held no real significance, from either an operational or overall cost standpoint. It had served to be the one thing he and Patricia admittedly disagreed on. But no amount of discussion on his part had worked to change her mind. In fact, their last conversation on the matter, only days earlier, had almost turned heated.

"Jim," she had said, "I'm not going to bend on this one, so forget about it. Barring required it's union officials to hold down production jobs and if they can, we can."

"But, Pat," as she had encouraged Jim and others to call her, "the facts are we're fighting a long-standing precedent. By your own admission, it was in place from the beginning at Barring and not something they won at the table," he responded.

"So what?" she challenged.

"So, we will have a difficult, if not downright impossible, time achieving it. I would think you could see that," he quickly replied.

"What I can see, Jim, is that we're at an impasse. When that happens it's my job to reach a decision and your job to carry it out," she said, smiling to emphasize she wasn't upset but was making a point.

"I get the message. But I think it's also my job to tell you when I see a problem developing and I believe this has strike written all over it," he said as he stood to leave.

"You're right about keeping me informed. But a strike? I don't think so. They're smarter than that. They know we have a strike hedge and, as you're aware, we have inventory on hand to cover our requirements for up to 90 days, if need be. There's no way they are going to keep their members out on the street for three months," she responded.

"I'm not so sure. They've done it before. Back in '85, they went out 13 weeks and it was a bitter strike that left some lasting wounds, according to what I've heard."

"That was a long time ago and things have changed," she noted, pausing a moment. "Now, I don't know how to say this any plainer than I already have, but once again I'm determined to see this through. So, let's just drop the subject. All right?"

"But it is my plant we're talking about here and," Jim stubbornly continued before she raised a hand to silence him.

"We're not talking about just your plant, Jim. We're talking about Denning as a whole. We intend to see this accomplished at each of our manufacturing facilities."

"Would you at least let me throw something out for your consideration?"

Jim could see her irritation. He realized he was testing her patience, but he felt he could not walk away from it since it was too important.

Patricia let a long breath escape. "All right. Have at it."

"Thanks, I appreciate your patience on this, but it is important. What I've been thinking about was a way to accomplish most of what you're looking for and still provide an out for the union. What if we changed our demand to include everyone except for the union president? This way, we get what we want for the majority of those involved and they still have their primary representative, the union president, free to conduct business on a full-time basis."

She thought about his words. "I'll admit that's interesting. Let me think it over and discuss it with Norm," she replied.

Jim had left feeling relieved he had got his idea out and felt good that Owens hadn't summarily rejected it. It appeared she would mull it over. As Jim sat in his office and Phil brought him up to date on the day's events, he reflected on his conversation with Patricia.

"They stomped out twice today and said it was useless to continue. While we managed to get them back after they cooled down, we're going nowhere fast," noted Phil.

"How does Norm feel about it?" asked Jim. Norm was the company's chief negotiator.

"He thinks they may walk," replied Phil.

"I assume he's told that to Owens," responded Jim.

Phil shrugged his shoulders and lifted both palms upward in a gesture of uncertainty. "He doesn't tell me a lot about what he takes up the ladder," said Phil.

"I know. Norm plays it too close to the vest for me. I'll just have to touch base with Patricia and find out if she's up to speed," Jim replied.

"Well, tomorrow is a key day. We'll either come out of there with something they are willing to take to the membership or we'll know we have a strike on our hands."

"What's your gut feeling about it, Phil?"

"They're prepared to go to the mat, if necessary. Of course, they would have some convincing to do if they were to suggest to the membership they should walk over what isn't a pay and benefits issue. While they're not thrilled about the financial package that's been offered, that's really not their primary point of contention. It's about us insisting the officers hold down full-time production jobs. In my judgment, it's the perception it would create rather than any inability on their part to conduct business adequately. They aren't saying it, but what they are thinking is that it would show the members a serious decline in the union's power and influence. So, I think they'll hold fast on their position and if they do go down, they'll go down swinging," he replied.

"I agree with your assessment. I suggested something to Pat I hope she'll take to Norm," said Jim, telling Phil about his proposal and his reasoning. Phil agreed with Jim's idea and told Jim he had already mentioned it to Norm.

"As they say, good minds tend to drift in the same channels. Believe it or not, it's an option Norm and I discussed during one of our breakout sessions only a couple of days ago. Unfortunately, he said Owens and Brooks would never buy in. Apparently, they want to follow the Barring model on this one," Phil responded.

"I like your comment from before, but I had never heard it put quite that way. 'Good minds tend to drift in the same channels,' wasn't it?"

Phil smiled broadly and nodded.

"I've got to tell you, Phil, I like that!" Jim complimented. They both shared a friendly chuckle before continuing. "But regarding Owens refusing to buy in, I'm not so sure about that. I believe I got her attention, so we'll have to wait and see," noted Jim.

WARRING SCORECARD, PART 4

As Plant Manager

I would have to give Jim an A-. Not perfect but clearly better.

He displayed some impressive growth and maturity in how he attacked the expenditure problem. In doing so, he pulled his team together and came up with a means to overcome the problem within the same month and potentially for the year. When some extensive pressure was on, he bent but refused to break when he knew he was right. He did this by pointing out that the major problem was an obsolescence issue created because of a customer demand and not because inventory had been unnecessarily stockpiled.

It is anyone's guess when it comes to the matter of Jim's personal decision to decline Harold Jenkin's offer, but he did show some wisdom in understanding he was still inexperienced in his role. Therefore, to venture out with counseling and directing others in what they should be doing—before he had first proven himself fully capable as a plant manager—would have been risky. On the other hand, I would be first to admit such opportunities surface on rare occasions during one's career. Therefore, I cannot offer a right or wrong answer. Such a decision is a personal one, based on factors that can't be neatly wrapped into a do or don't package. However, as you will later see, the door hasn't yet completely closed for Jim.

Jim also did a good job of being persistent enough to give his boss his idea on how to resolve the developing union and company standoff. Jim could have more easily gone with the flow, so to speak. If the result is indeed a strike, he can take what will be a hard position to challenge, which is it wasn't his idea and therefore something he had no control over. Instead, Jim had enough fortitude to stand up for the overall benefit of both the company, as well as the union, and for that, he should be commended. I want to stress again that a plant manager has as much obligation to create a good working environment for the established union, as he does for making the manufacturing sector a total success. The two go hand in hand.

In the upcoming chapter, I will address the unique role a plant manager must play in being a solid ambassador for the company and its employees, and whether a union does or doesn't exist is irrelevant. If one is in place, the plant manager must help make it as successful as possible. I realize that statement may shock some, but I'll support my reasoning and logic later, so please bear with me. All in all, Jim is steadily improving as a plant manager.

As the Chief Conductor for Lean Engineering

Sadly, his efforts here are still short of the mark. Again, when dealing with the job's common problems and specifically upcoming contract negotiations, sizable scrap, and operational cost problems—along with other distracting personal issues—Jim has allowed himself to sink into a mire of complacency. He has allowed the job to drive him rather than vice versa. Though he can reassure himself he is an effective problem solver, which is certainly an ongoing function of a plant manager, he has forgotten the bigger responsibility he holds. That, specifically, is to manage for the future. But, as you will see, this is about to change.

6

Dealing with the Unspoken Duties

THERE ARE ASPECTS OF A PLANT MANAGER'S JOB that are not commonly discussed or adequately addressed. It is important to do so and I will start with some statements and then return to detail them. However, before doing that, I should set the stage.

You could view some of the following statements as controversial because they typically go against the grain of conventionality. That, of course, is the way we have always felt we should think and act even though we haven't specifically been told to do so. I'm referring to those convictions that are never openly debated and/or company policies which have never been clearly established. On the other hand, these unspoken rules can sometimes be terrible stumbling blocks to an operation's long-term success.

Here are a few examples:

- If you're a member of the management team, you should say or do nothing that could be construed as "supporting" a labor union.
- If you have highly effective employees, your duty is to do everything within your power to keep them there. If necessary, this can include tactfully blocking a promotion to another branch or division within the company.
- Even though clearly established policy supposedly applies to everyone, certain exceptions must be taken into consideration to do what's best for the business.
- You must never lie to your customers, but you are not obligated to tell them more than is necessary to keep them satisfied.

These unspoken rules exist for one reason or another even in the finest organizations. Though they are not generally discussed, they are powerful rules that come as a result of applied perception. I'm referring to those strong convictions that managers and employees feel are expected of them. In truth, these may not represent the company's actual position and, in fact, could be just the opposite. These are commonly left to people's judgment by most companies because they aren't something comfortable to address and establishing firm policy has not easy solutions.

Allow me to take one example and expand it to demonstrate my point. Right or wrong, most companies prefer a non-union environment. Therefore, they usually do what they can (within the limits of the law) to fight the establishment of a labor union. However, imagine a company putting together a written policy that said anyone who was a part of the management team was forbidden to say or do anything that might be viewed as support for a union. First of all, it would probably be too vague to enforce. Second, it could lead to a serious conflict between the company and its employees because, by law, a company must avoid willfully coercing or intimidating its employees against either joining or remaining in a union. As a result, the unwritten rules then become how employees (who are part of management) believe the company would prefer they act, react, interface, and/or respond. Many times, the results may in no way reflect the true feelings of a company's senior management. Regardless, it is generally viewed by those who belong to the union as the company's position.

So, what is the solution to this perplexing issue, as it applies to running a manufacturing operation? In my judgment, it centers on plant managers and how they address such matters. Remember, you may view some of the statements below as irregular (if not, downright insane). But, be patient and read on because the primary purpose of this work is to help those starting out as managers. What follows can assist you by providing a road map for being effective in a vague area, typically explored only through trial and error.

- Plant managers must feel as obligated to the success of an established union as they are to the goals and objectives set forth by the company.
- Plant managers must assume a standing obligation to assist those employees (who have supported and helped the company) to better

their careers, even if this creates problems for the operation by encouraging them to leave.

- Plant managers must enforce the company's rules and policies by showing no special favoritism to any employee, under any circumstances.
- Plant managers must insist that when problems arise affecting the delivery and/or quality of products, that customers must be given the whole truth and nothing but the truth, without exception.
- Plant managers must, upon discovery, ensure employees immediately know about layoffs or other situations that would force manpower reductions.

I am certain these statements have your attention. In fact, some of you are wondering if I haven't gone off the deep end because what I'm suggesting is atypical thinking. As I will show you, this is how all good plant managers should do their job. Therefore, I will take each of these in order and provide you with my logic on the subject.

THE DUTY TO ASSIST IN THE SUCCESS OF AN ESTABLISHED UNION

If a union already exists, it is to the benefit of the company and the particular function over which the plant manager reigns, that the union be as successful as possible.

While I shouldn't have to say this, I will try to avoid misunderstanding: I am in no way implying plant managers should go out of their way to support establishing a union. That would, if anything, be frowned on just as much by the union as the company. Further, it would serve no logical purpose and would be professional suicide.

What I am speaking about, however, is doing all one can to create an environment where the union can operate without being made to look like the number one enemy. In fact, the ideal working relationship would be a partnership with the union designed to address and resolve problems, which make the company less competitive. (See *Fast Track to Waste-Free Manufacturing*.)

An inadequate relationship with a union can be tremendously disruptive. I know because I have worked in such an environment. If you feel you could live with that, consider the horrible influence this kind of relationship can have on teamwork and results. I can assure you it will be a boulder in the path of progress. I can also assure you that productivity, customer satisfaction, and the overall quality of work life will suffer.

Some would challenge this, pointing out that when efforts are made to build a compatible relationship, the union's elected officials will not cooperate because they have become hardened against management. I would simply ask: Why do they feel this way?

The facts are, during all my years in manufacturing and management, every situation I have seen with a bitter relationship was because of poor management strategy. If that sounds as if I am placing the primary blame on management, rather than the union, that is correct.

If a company fairly and consistently deals with the union, how long can union officials protest (and get the membership to buy in) the company has no intention of doing the right thing? The answer would most often be not long. I emphasize most often because turning around a sour working relationship with the union is difficult and there can be cases where the damage is so severe that correcting it is next to impossible. However, even under the worst scenarios, you can steadily improve the relationship if you apply consistent and fair management practices.

If an established union cannot function in the eyes of its membership, the operation's success will decrease because of mistrust, hard feelings, and disruption. No exceptions. The key to resolving this rests with the plant manager assuring dealings with the union are, at all times, fair and unquestionable. When put into practice, goodwill and employee confidence will steadily improve.

THE DUTY TO ASSIST EMPLOYEES IN BETTERING THEIR CAREERS

Over my years in manufacturing, I saw repeated cases of good employees being held back from promotions or moves that would have improved their career. Conversely, I saw situations of employees (who were not the best of performers) being actively sponsored for jobs in other functions that enhanced their future.

You would think the opposite situation would exist. You would expect a high-level performer—the person who goes above and beyond the call—to be the first to reap the rewards of a career advancement opportunity. This is often not the case. Why? Primarily because of fear and poor management practices.

The fear I'm speaking about is a concern (often genuine) that talent will not be available to replace an individual. The poor management practice I'm referring to is the lack of an adequate process that effectively addresses and deals with this matter. In most cases, this is created on one end by management not doing enough to recruit and train replacement talent and on the other by poor performance tracking and documentation.

When a good employee is intentionally held back, it is never due to an intent to hurt or punish this person. As I've said, it is due to fear. That fear is that without the employee, the performance of the operation could be hampered, in some cases, dramatically. You often hear the following reasoning:

- John is the only one who knows how the equipment really works.
- Jane is the only person we have who really understands how to do the job.
- The customers would protest if we pulled Kyle off this job.
- We can't get the project completed on time without Lucy's help.

What they are really saying is: "We haven't taken the time to consider the possibility of losing John, Jane, Kyle, or Lucy and have no contingency plans in place."

I would be the first to tell any plant manager that if you have a person in your operation who is irreplaceable, you have yourself out on a weak limb. When this happens, unless you are lucky, you will face one of two major problems. The first is your operation would be in trouble should that person decide to leave the company or become ill, injured, or otherwise incapable of doing the job. The second is that you're going to disappoint a good talent because you have held him or her back from a career enhancement opportunity. Either way, you lose, the employee loses, and the company loses.

Often, an opening initiates advancement opportunities. Therefore, the chance of delaying someone from moving into the position, until such time their replacement has been recruited, hired, and trained, would be viewed as impractical by those actively seeking to fill the position. In most instances, personnel will tell you it can't wait that long and recruit someone from outside the company. The opportunity for an internal advancement is then lost.

On the other hand, the statement *everyone is replaceable* is still true. In most cases, more than one person can do a job. Could losing this person potentially create some hardship on others for a time? Certainly. Could it potentially cause some minor problems and difficulties? Certainly. Could it keep the operation from running and satisfying customer demand? Almost never.

When situations do arise where an employee cannot be immediately released for a new assignment, plant managers should take some special steps:

- Work diligently to provide the person with some special compensation.
- Insure someone is trained as a capable backup as quickly as possible.
- Finally, establish a process which will assure that no worthy employee will be held back from a career-enhancement opportunity.

THE DUTY TO ENFORCE THE RULES CONSISTENTLY

On a few occasions during my plant manager career, it was requested I show favoritism to certain individuals who had broken company policy. In every case, special circumstances existed which could have lent credence to showing some special consideration. The most compelling of these involved employees who had been good performers and who had no previous violations on their record.

In two of the cases, the employees had been with the company for years and were approaching retirement. I knew that losing their jobs could be a devastating experience in countless ways. If nothing else, they were approaching an age where even getting consideration for another job would be difficult. However, after much deliberation and some sleepless nights, I showed no favoritism. On the surface that probably makes me sound a little cruel and uncaring, but nothing could be further from the truth. To this day, I think about those decisions and find myself fretting about the lasting effect they had on the employees' families. But if I had it to do over again, I would make the same decisions.

In one of the more compelling cases, a long-time employee had been accused of taking some rather inexpensive building materials from one of the plant's storerooms. Of course, stealing from the company was a termination offense. As the story unfolded, however, we discovered he had used the material to help put the finishing touches on a modest house he was building for his daughter. When all the facts surfaced, we realized he had done this because he had run into some major financial problems, the result of his wife's lingering illness. In addition, his daughter had just divorced and was struggling financially to support three small children. Because he and his wife's home simply wasn't large enough to accommodate everyone, he felt compelled to build housing for his daughter and her children quickly. Later when approached about the matter, he admitted he had taken the goods, but he swore he planned to replace them as soon as he was financially able. To this day, I honestly believe that was his full intention. However, the decision was ultimately mine and I decided to let him go. Here are the reasons why:

1. The company policy for such a violation was termination. Period. However, if I had supported a special exception for this person based on his compelling plight at the time, I would have set a dangerous precedent. From that point forward, any employee found guilty of stealing would have had the right to appeal termination. You have to wonder how long it would have been before someone would have used the excuse, "The devil made me do it."
2. You must base such decisions on the best interest of all employees. Being a plant manager and making appropriate decisions isn't always fun. As a result, even when you are right, you won't always feel good about it. In fact, you will be required to make some decisions during your career whose consequences may haunt you for years to come. However, the positive aspects of the job far outweigh the negative, and if you have done a good job, you will find you've help far more people than you've hindered.

To some extent, the job of plant manager is similar to that of a judge. Both spend the majority of their time seeking out the facts and making decisions. The plant manager has policy to guide him in his decision making. The judge has laws. The common denominator for both is precedents, which once established are as binding as any written word.

Therefore, the plant manager's duty is to insure the company rules and policies are enforced equally for all and with favoritism toward none. It's that simple and, unfortunately, an often unappealing responsibility. But it is absolutely necessary.

THE DUTY TO BE HONEST WITH CUSTOMERS

One of the more horrible mistakes a plant manager can make is to influence an employee or group of employees to be less than truthful with customers. There are two rules that should apply. First, you must encourage your employees to tell customers the truth and nothing but the truth, even when it hurts. Second, you must practice what you preach.

Though this is admirable, some would contend it could be taking a good thing too far. Even though we may prefer to avoid substantial untruths (the *big* lie), we sometimes feel telling customers less than the entire truth (the *little* lie) is all right, if not absolutely appropriate. Wouldn't you agree, however, the problem becomes defining big and little?

Customers may not know or care about how you would go about defining a big versus a little lie, but they know one when they see one and they don't like it. In fact, one sure way of losing customers is to have them believe they can't trust you to tell the truth. I can assure you, unless you are the only game in town, they will only be around long enough to find and establish your replacement. Therefore, logic dictates you shouldn't give your customers the excuse to leave, and to accomplish this, be honest and straightforward with them.

For those who feel telling less than the truth is acceptable, I feel this has been the biggest factor contributing to the practice of direct customer involvement in running a supplier's business. The automotive industry is notorious for this, and for good reason. Over the years, half-truths by suppliers led to serious and costly recalls for auto makers. In an effort to avoid this, the Big Three insisted that certain common initiatives (e.g., ISO 9000) be undertaken by all suppliers and have further held the right to direct how certain manufacturing processes be constructed, run, and maintained. For the most part, this has worked. There have been far fewer field problems and the overall dependability of automobiles has substantially improved. But, serious problems have arisen in providing a conducive environment for fast and effective process improvements (by suppliers) and, thus, increased productivity and lower overall costs. This is because a serious level of suspicion and mistrust has continued that would not have been necessary had the full truth always been told.

Therefore, it is the plant manager's duty to ensure customers are provided the full story when issues arise.

THE DUTY TO INFORM EMPLOYEES

All plant managers enjoy standing before their employees and communicating good news. Why wouldn't they? Everyone feels better. On the other hand, many feel delaying bad news as long as possible is the way to conduct business. Their reasoning is that bad news stirs up the workforce and the resulting impact will often be lower throughput and/or productivity. For example, many take the position that informing employees of a pending layoff would create a longer period of disruption as opposed to delaying this news until deemed necessary.

While there is some truth to this, what should come to mind is an even more damaging potential, which pertains to the confidence the workforce holds in management to be forthright with them. If employees feel the company will keep them informed (be it good or bad news), they will more often be satisfied employees. If the opposite is true, they will more often be disgruntled and suspicious. If you allow the latter to go on long enough, employees can become rebellious and difficult and, as a result, almost every facet of the business can suffer.

I speak from experience on this matter, and here I probably established one of my better overall reputations. If I ever did anything right from the offset as a plant manager, it was always being a straight shooter when keeping employees informed. My position on this was simple and straightforward: If you know about news employees must eventually know, tell them NOW and not later.

Anyone would be foolish to believe that if great care is taken, bad news will not leak out before intended. Believe me, bad news—whether it's a pending reality or a remote possibility—will always leak to the workforce. What is worse is when something like this leaks to the outside and employees learn of it through secondhand rumors. What I am suggesting goes even further. If a false rumor starts to circulate, which is bad news, you must let the troops know the rumor is inaccurate. However, a word of clarification is possibly warranted because I am referring to job security rumors. This would include such things as false rumors about potential layoffs and/or workforce reductions, which can come in many forms and for many reasons.

In my judgment, a plant manager must provide an environment where employees feel that they will always get important news, and regardless of how good or bad that news might be, they will always be the first to know.

THE WARRING ADVENTURE, PART 5

Patricia Owens and Phillip Brooks met with Jim and collectively heard him out on his suggestion. On the bright side, they didn't give it a thumbs down. However, they made it clear they were going to have think hard about it before they would be willing to concede to something less than they had originally hoped to accomplish.

In the factory, tensions were starting to develop with a few minor incidents of equipment sabotage. Though common during contract negotiations, Jim found them annoying. At the moment, however, all that was forgotten as he answered his phone, only to find found Jason Andrew on the other end of the line.

"Well, how have things been with you?" Jason asked.

"Just fine. And you?" replied Jim, surprised by the call.

"Not the best, my friend. We're having some start-up problems with Miller and as a matter of fact that's why I'm calling you." Miller Industries was the company that had taken the business away from Denning. There had been some predictions in marketing that Miller wasn't capable of keeping up with Aklin's demand, but Jim and others had excused this as a sour reaction to losing the contract.

"I'm sorry to hear that," Jim politely replied.

"Thanks," Jason said.

"I know it's unusual for someone in my position to be making the initial contact on something like this, but since you and I developed a good working relationship, I agreed to give you a call and feel things out. What it boils down to, Jim, is that we're having some big problems at the moment keeping up with customer requirements. Miller overextended themselves and can't keep up with demand. They're correcting the problem, but in the meantime we're in hot water with some of our customers. So, I'm coming to you with hat in hand and was wondering if you would be interested in helping us out?"

"Certainly, if at all possible." Jim responded.

"The truth is, if it had been up to me, you'd still have the contract. But as you know, manufacturing doesn't call the shots on those matters. Miller makes a good product and its quality has always been outstanding. But it has a history of delivery problems and even with all the promises we were made, nothing has really changed from that standpoint. The truth is that they're meeting their basic agreement with us, in terms of volume, but in this business there are always fluctuations in demand and we have to respond to those. Miller has taken the position that it's not their fault, but ours for not predicting the volume properly and building it into the contract. Frankly, a war is going on at the moment between our purchasing group and Miller."

"I see," Jim replied. "So how could we help?"

"We have an extra set of cutting dies for these products and were wondering if we gave you most of the added volume we're experiencing if you'd be interested? I realize you're not in a position to make a solid commitment for Denning at this point. I just wanted to find out how you'd feel about it."

Jim knew he had to take some added precaution in how he responded. If Jason went back and announced Jim had jumped at the chance, it might go against the grain of sales and marketing's approach. They might not want anyone in Aklin to believe Denning was interested in some small piece of the business, much less to pitch in and help them with a problem created by one of Denning's leading competitors. On the other hand, Jim wanted to leave an impression he was genuinely interested. Therefore, he carefully chose his words.

"I'll certainly take your request forward and let you know the response I get. Is that fair enough?"

"That'll be fine. When can I expect to hear from you?" asked Jason.

"Tomorrow at the latest," Jim promised.

Following this, Jim beat a path to Patricia Owens' door. He related his conversation with Jason and his request, then waited for her response.

"If he's asking us to help resolve a problem created by a company who took the business away from us, the answer is no. Not unless Aklin is willing to give it back lock, stock, and barrel," she stated.

Jim was surprised. "We know that won't happen. They have a binding contract with Miller at the moment."

"Then as I've said, the answer is no," she flatly stated.

"Pat, I respectfully disagree with the position you're taking."

"Dare I ask why?" she responded, appearing somewhat irritated at his persistence.

"Don't take what I'm saying the wrong way. It's just that I see a number of good reasons to help them out if we can," said Jim.

"And what would that be?"

"First, it can only help when it comes to the potential of getting a new contract in the future. And we can't forget Aklin is the largest customer we've ever had in terms of volume. Last, but not least, it could help our negotiation efforts. The union's taken the position we've lost Aklin forever. This would serve to show we haven't," replied Jim.

"Well, I could make an argument about that. Though Aklin was the largest in terms of volume, they haven't been a good money maker for us. If the truth be known, they didn't go to Miller because of poor quality on our part. They went because Miller undercut us on price. And regarding the union, I see where they could use this in an opposite manner. Knowing Aklin has come back to us could make them feel everything was all right to begin with and what we're asking for to improve the business will not be necessary. You can look at this many ways, Jim," she contested.

"I suppose so," he admitted, "but you can't argue it would increase our chances of a new contract with Aklin in the future."

"I give you that much. All right. Let me run this by a couple of people and get their thoughts."

"I told Jason I'd get back to him tomorrow at the latest," Jim noted.

Patricia studied him for a moment, then smiled broadly. "You're a persistent cuss, aren't you?"

Jim returned the smile as if to say he agreed.

The following day it was agreed by all that Jim would call Jason Andrew and let him know Denning would look at the work. He was told to suggest that someone in Aklin's purchasing department contact Denning's marketing group so it could determine the company's volume, pricing, etc.

Jim did as he was asked, Jason responded positively and the process was formally initiated for a wait and see situation. Jim got other pleasing news when Patricia called him aside after the meeting to inform him they were going with his suggestion on the union issue.

Late that afternoon, Phil Tanner dropped by Jim's office to inform him the union was intrigued by the proposal. In fact, he was convinced that although they expressed disappointment at the suggestion, they were going to strongly consider it because they could still view it as a win. Plus, they had no doubt the company would go to the wall to achieve some solid concessions in this area. Therefore, Phil was convinced that after some posturing and complaining they would buy into the proposal and use this as leverage for other things they wanted to accomplish.

By the next morning, the news of the company's proposal had indeed leaked out to the general plant population. It was a conscious ploy on the part of the union negotiating committee to measure the response. Various feedback channels indicated the general membership really wasn't that concerned about whether or not union officials should hold down full-time production positions, and clearly wages and benefits were going to move them to vote yes or no.

For the first time in a while, Jim was feeling more comfortable. The contract had a good chance of being signed, and he was almost certain he would have good news to give the troops regarding the Aklin business.

Then, as if lightning had descended out of a clear blue sky, it happened and Jim couldn't have been more shocked or surprised.

He was returning from Patricia Owens' office and had stopped in one of the main production aisles to chat with Maynard Williams, a heavy set, jovial maintenance man with a reputation for a bright sense of humor. Everyone liked Maynard.

The concussion that interrupted their conversation shook the entire building. Everyone was looking around at each other and Maynard, who had been facing Jim and the east end of the plant, was the first to react. What Jim noticed most—something forever burned into his memory—was the way Maynard's eyes had widened with astonishment and fear.

"My god!" Maynard exclaimed and then without warning bolted forward and around him, his shoulder catching Jim's and almost knocking him off his feet. But Maynard didn't bothered to stop and apologize.

As Jim recovered his balance and turned to look, he saw Maynard hurrying down the aisle with one unbuttoned strap on his overalls flapping wildly behind him. He was scurrying toward a set of double doors that separated the main facility from what was known as the chem lab.

A huge ball of smoke began to billow out the doors and almost instantaneously, a series of fire alarms kicked in and began to pulsate. Suddenly, people were scampering everywhere and Jim found himself thinking that someone must have planted a bomb. Then just as quickly he told himself it couldn't be. People make bomb threats. They don't actually do it!

He quickly let all that pass, however, and followed Maynard, who had now entered the lab area and disappeared into the hazy, lingering smoke. Jim was following at a trot and was almost through the doors when a second and more powerful explosion erupted. He was thrown off his feet and landed squarely on his back. In the process his head slammed against the floor and he found himself fighting consciousness. An instant later another series of sprinklers were activated and the next thing he knew he was being drenched. For an instant, he feared he was going to drown, but gathered himself and struggled to his feet, with the force of the sprinklers pounding down around him.

He stumbled through the doors, ignoring the pain in his right arm and a burning sensation on his face. Inside the sprinklers had done their job and any fire that had accompanied the blast had been extinguished. However, heavy smoke filled the air and he was finding it difficult to breathe. As he worked his way forward he stumbled over something, but managed to break his fall with the palms of his hands. In trying to get back to his feet, he realized it was Maynard.

Instinct told Jim the first thing he needed to do was get the man out, but as he struggled to pull him to his feet he discovered Maynard was too heavy. Throwing caution to the wind, Jim grabbed Maynard's overalls in both hands and dragged him along the floor until they were back through the doors and past the water pouring from the sprinkler system and saturating the floor.

When he finally stopped, he knelt to look at Maynard and his heart sank. The skin on Maynard's arms and face had been scorched from the heat of the blast and although Jim could see he was breathing, it was labored. Feeling a hand on his shoulder, he turned to see a hooded face. It took him a moment to realize it was someone in a firefighter's outfit.

"We'll take over here," he heard a voice say as he was pulled to his feet. "Are you all right?"

Holding Jim at arm's length by the shoulders, the firefighter studied him up and down. "We need to get you an ambulance." Suddenly, Jim began to feel faint and his knees buckled. The last thing he remembered before he slipped into unconsciousness was the voice becoming fainter and fainter as it shouted for help.

On the positive side, no one had been in the lab at the time of the initial blast. Employees who worked in the area had just vacated the premises only minutes before to take a scheduled break. Unfortunately, Maynard hadn't known that and his son, Vincent, worked in the lab. Instinctively, he had run to help and Jim had followed; as a result, they were the only employees injured in the blast.

Jim had suffered a fractured elbow and some slight burns to his face. Other than being sore for a few days and having to wear a cast on his arm, he quickly recovered. Maynard on the other hand ended up fighting for his life. Along with second- and third-degree burns on both arms and a portion of his chest and head, he had inhaled enough superheated air from the second blast to singe both lungs. He came down with pneumonia and his condition had been listed as grave. But being an extremely strong and robust man, he made a remarkable recovery within a month.

It had taken a few days to establish the cause of the explosions. Trichlorethylene used to clean a number of parts and components was being pumped from a tanker just outside the building into a series of holding tanks inside the lab. Because the driver failed to ground one of two hoses used to discharge the fluid, a static charge was created and proceeded to ignite one of the holding tanks. After the initial explosion, fire and heat had then set off two of the other holding tanks. The result was what appeared to be one additional and more powerful blast, which in fact was two separate explosions that occurred almost instantaneously.

OSHA quickly investigated the accident and since a chemical was involved, the EPA also visited the site. Fortunately for Denning, absolutely no violations were found or cited. As a precaution against a future incident, however, a new procedure was established, which stated a Denning employee must accompany the tanker driver when connecting grounding devices. In addition, the driver had to sign a form, prior to pumping any fluids, indicating that he or she had followed all necessary grounding procedures.

The incident happened on a Friday, and after the weekend to recoup, Jim was back at work the following Monday although he was limited in getting around. Ginger wanted him to stay home for a few days, but it was to no avail. He felt that given the conditions, he had to be there if at all possible.

In the local newspaper, he'd been portrayed as a hero and was given credit for saving Maynard's life, at a substantial risk to his own. Jim explained he had followed Maynard into the lab to see what was going on and not out of some fear for Maynard's life, and that he had simply stumbled upon him after the second explosion. But heroics make for good headlines, so heroics it would be, regardless of how Jim tried to play the matter down.

Within a few days everything was almost back to normal, with the exception of the chem lab. Because the lab had been destroyed, on-site work had to be outsourced until the lab could be repaired. Jim and others worked feverishly on accomplishing this and with just enough inventory on hand to meet established schedules, they managed to not disappoint any customers.

Shortly following this some good news came Jim's way. Union and company representatives reached a tentative agreement on a new three-year contract. Two days later, the general membership ratified the vote. If anything, the incident had served as a peacemaker. The magnitude of it paled anything on either side of the table, and the mood shifted from an almost mindless debate to resolving the business at hand. As a result, the negotiations took on an atmosphere more conducive to making progress.

Then to top off what seemed to be a run of better luck, Denning signed, sealed, and delivered a contract with Aklin. Within days, plant personnel set up a temporary process for the work and brought back a number of people who had been laid off.

Some six weeks after the accident, Jim was sitting in his office when the phone rang. He waited for Linda to answer, but after the second ring decided she must have stepped out for a moment and proceeded to take the call himself. It was Patricia Owens.

"Hi Jim. I have Matthew Goens with me and I was wondering if you could possibly drop by for a few minutes?" Goens was chief counselor for Denning and supervised a group of younger attorneys who handled legal matters for the company.

A few minutes later he settled into a chair in Patricia's office and waited for them to finish a feverish conversation. Patricia got right to the point.

"Jim, Matt has informed me that a lawsuit has been filed by Maynard William's family. I'll let him give you the details, but I've asked you here on short notice because we wanted to bring you up to date on this," Patricia noted.

"Thanks," Jim replied, interested in hearing more.

Matthew gave Jim some news he certainly wasn't expecting. According to him, Maynard and his family were collectively suing the driver of the tanker, his company (Warthorn Chemicals), and the original equipment maker of the grounding devices involved. Along with that, Denning Corporation and specifically Jim, himself, were listed as co-defendants.

"Co-defendants?" Jim responded.

"That's correct. It's a bit complicated, but it reduces to a form of punitive damages. Though they can name you or any other employee they like, when it comes to something like this Denning assumes full responsibility, not that we believe there is any liability to begin with," Matt explained.

"I'm sorry, but I still don't understand," Jim replied.

"Look at it like this, Jim. Maynard Williams suffered a serious injury while he was at work, on the job. In fact, life threatening. You were the closest person in the senior management ranks when it happened and it came as a result of nothing he did, other than being in the wrong place at the wrong time. Give that much to any decent, self-respecting liability lawyer and you've got a case he or she would be chomping at the bit to take. However, all I've been provided up to this point is notification that a suit has indeed been filed. So, we'll just have to wait and see what they want."

"What would you speculate?" Jim asked.

"Well, as I was telling Pat, in all likelihood they're going to say the company didn't have proper safety standards and precautions in place. That, of course, would be a rather weak case since OSHA and EPA have given us a clean bill of health. In specifically naming you, however, it appears they have something else up their sleeve and it's difficult for me to speculate what that could be. Frankly, we just don't know at this point. But let me ask, did you say anything at the time of the accident to Maynard or anyone else? I'm interested in knowing because sometimes even an innocent remark, at a time like that, can be twisted around to fit some scenario they're trying to mold."

"Nothing that I'm aware of," Jim replied.

"All right, we'll just have to wait and see what develops," Matt noted.

Three weeks later, Jim was sitting in Matthew Goens' office providing a deposition, at the request of Quency Phillips, the lawyer representing Maynard and his family. In attendance was a recorder, Quency, Matthew Goen, and Jim. Quency had a not so glorious reputation as the city of McCray's chief ambulance chaser, but he was highly respected as a man who got the job done. After some preliminary questions on his part for the record, he began.

"Mr. Warring, you were talking with Maynard Williams in one of the main aisles in the plant at the time of the accident, correct?"

"That's correct," Jim replied.

"And after the first explosion occurred, both you and Maynard proceeded to the lab?"

"He went first and I followed shortly thereafter."

"Elaborate further on that if you will," Phillips countered.

"As you said we were chatting together and I was standing with my back to lab. After the initial explosion, Maynard who was facing that way immediately pushed by me and headed for the lab. I followed him a moment later."

"Did you say or do anything to stop him?"

"No, but he was gone before I knew it and probably couldn't have heard me anyway above all the machine noise," Jim responded.

"But you didn't try, did you?"

"Well, it really didn't enter my mind at the time."

"You saw an employee of yours heading toward what was obviously a dangerous situation, and it didn't enter your mind to try and stop him?"

"I didn't realize there was further danger. A second explosion was the furthest thing from my mind," Jim replied.

"There was already danger enough from the first explosion, wouldn't you say?"

Matthew Goens interrupted Jim before he could respond.

"You're not required to answer that," he stated calmly.

Phillips made no argument and quickly proceeded with another question.

"Mr. Warring, did you not clearly see Maynard Williams running toward the lab where an explosion had just occurred?"

"Yes."

"And you didn't think that as being a potentially dangerous situation?"

"Of course, but I didn't know he was going to enter the lab."

"What did you think he was going to do?"

"Again, you don't have to answer that," Matt interjected. He then asked that the official proceedings be halted for a moment. The recorder leaned back from her machine and placed her hands on her lap while Goens and Phillips conferred on the matter.

"Come on, Quency. He's not required to speculate on what another person was thinking and I'm not going to allow him to respond to leading questions like that. Let's just stick to getting the facts. All right?" Matt said.

"I want to establish what was going on in Mr. Warring's mind at the time. There's nothing irregular about that," responded Phillips.

"That's correct, but he answered that question. He said a second explosion was the furthest thing from his mind at the time. Now you're beginning to push the envelope a bit."

"All right. I've got what I need for the moment, so let's just continue," Phillips conceded and with that the official proceedings started again.

"Mr. Warring, you didn't make any effort to stop Maynard Williams from entering the lab, did you?"

"I guess not," Jim responded.

"Please respond with a yes or no. Did you or didn't you?"

Jim looked at Matt who nodded his head as a sign for him to do as he was asked.

"No."

"Then why didn't you stop him?"

"I suppose I didn't see a real danger to him at the time."

"Thank you," Phillips responded. It was obvious he had finally heard what he was looking for.

"Now, Mr. Warring, when you followed Maynard into the lab could you explain the conditions?"

"Conditions?" Jim asked.

"Yes, what were the conditions inside the lab at the time?"

"Well there was smoke but no fire that I could see."

"Heavy smoke?"

"Yes, quite heavy."

"Heavy enough that you could hardly see?"

"That's correct."

"Heavy enough that it was hard to breathe?"

"As I recall, yes."

"Could you see smoke coming out of the doors leading into the lab before Maynard entered?"

"Yes."

"And given all this you still didn't see this as a potential danger?"

"As I think I've said, I had no idea what he was going to do," replied Jim.

"Well, wasn't it more than obvious to you—"

"Enough," Matt interrupted. "As far as I'm concerned, unless we get off this particular line of questioning, this deposition is over," he continued.

Phillips smiled politely as he raised both hands into the air in a display of frustrated consent.

"No problem. I've heard enough," he said.

Three weeks later Jim, Patricia, and Matthew gathered to go over the official lawsuit filed by Maynard and his family. In essence, the suit charged that Denning had failed to have adequate safety procedures in place and that Jim, as plant manager, had failed to recognize an obvious danger and stop Maynard from proceeding into harm's way. More specifically, it charged that Denning was neglectful by not requiring the driver of any tanker to go through a written checklist before proceeding to discharge a volatile chemical.

"So, what do you think?" Patricia asked Matt.

"Well, as I've already told Brooks, after reviewing it with my colleagues we collectively recommend the company seek a no-fault out of court settlement, which we're reasonably certain we could achieve. While we believe it's a weak case, overall, the question becomes whether or not it's something you really want to fight, because a lot of negative publicity will come along with it. As you know in situations like this it isn't always what's absolutely right or wrong, but rather what's the most expedient and least expensive way to resolve the matter," Matt replied.

"And what did Brooks have to say?" asked Patricia.

"He agreed but wanted to get your input before we proceed in that direction."

Jim sat patiently and listened to Matt and Patricia discuss the matter. In his mind he felt the company should fight the charges. On the other hand, he wasn't a lawyer and it was his first experience with something like this. As a result he chose to listen and respond, if and when he was asked. But as the meeting proceeded, it became clear they weren't going to lean on him for professional advice. However, as Matt further explained during the discussion, they had used Jim as a convenient means to strengthen their point. In fact, he had said, if that was their entire case he wouldn't hesitate taking it to court. He then went on to state that for the record, he was certain they could get the personal charges against Jim dropped, which sufficed to make Jim feel somewhat better about Matt's proposal. With that, they agreed to follow Matt's recommendations and the meeting ended.

GETTING THE BEST OUT OF YOUR RELATIONSHIP WITH THE BOSS

During the course of your career, you will likely be required to report to different people with different personalities, who have many ways of leading and directing the activities of others. In some cases, you will thoroughly enjoy the experience. In others, you will not. In every case, you should get the best out of the relationship with your boss. Doing this properly requires concentration and not just responding to spontaneous and often thoughtless reactions that drive the relationship.

In my particular case, my career led me to work in a number of different locations and report to many different people. I had to learn the hard way about getting the best out of the relationship with my boss and in the process I made more than my share of mistakes. With time and experience comes some degree of wisdom and I therefore have some important opinions relating to this matter.

In the beginning, like most people, I felt the relationship with my boss was going to be what it was intended to be and that I had little, if any, control over it. How wrong I was. I discovered that the result of any professional relationship can be much more than some uncontrollable consequence. It can often be what you consciously decide to make it.

At some point you will be required to report to someone you don't like and/or respect. If this has already happened, you know how difficult it can be to maintain a professional and productive atmosphere. In fact, I have witnessed relationships where total disrespect and even bitter hatred set in on both sides. In this situation, little will get done effectively, and it is usually only a matter of time before something is forced to give.

This can happen in a number of ways. The subordinate becomes fed up and takes another job. The boss decides he has no choice but to replace the individual. Sometimes the matter is resolved by the boss moving to another position. Regardless, the fruitless time and energy expended can damage careers and does nothing for effectively accomplishing the job.

So, what are the rules of engagement when it comes to establishing and maintaining a satisfactory relationship with your boss?

1. Show the position its due respect
2. Keep judgmental attitudes out of the relationship
3. Be the first with the best continually

The following serves to explain these rules of engagement in more detail:

Though you are not required to like your boss, you should keep an appropriate level of respect for the position. If this seems somewhat military, the truth is we could all benefit by assuming such an approach. Doing this isn't some effort to provide a false sense of respect for the person involved. Rather, it is an effort to establish an environment where effective communications lead to achievable results. Most important, this approach defuses the creation of a defensive atmosphere on both sides of the equation. As the old saying goes, "It takes two to tango." Therefore, once a subordinate clearly shows he or she is not willing to dance, the boss doesn't have to worry about his footwork and can keep the relationship professional and constructive.

Remember that your boss had some distinct qualifications and/or professional accomplishments to win his/her current post; unless he or she happened to be a close relative of the owner and was given the position regardless of appropriate qualifications—which is unlikely. Therefore, always respond to what your boss has to say with an open mind and a non-judgmental attitude.

Consistently show your boss you will be the first with the best when it comes to responding to assignments and/or commitments. Once it has been agreed (or, at least, understood) you will take an assignment, then do it with the highest level of energy and responsiveness. Of all those involved, work to be the first in getting assignments completed and put forth the best effort possible. Even though you won't always be able to accomplish this level of achievement, keep the objective in mind. Do this and your boss will come to recognize and respect you as time goes by.

Practicing these rules of engagement will help establish a conducive relationship for effectively doing the job. However, what if you have a boss who is politically motivated, who is uncaring, who is self-centered, who represents the accomplishments of others as his or her own, or any combination of these and/or other less than appealing characteristics? Unfortunately, such people exist. How do you react to this? The answer is in the same manner as previously mentioned.

Look at it like this. There is nothing in the rules of engagement noted that is either illegal, immoral, or irresponsible. In fact, for the most part, these call for little more than a display of good manners and professional diplomacy. Becoming openly irritable or challenging never leads to anything constructive and, frankly, most often isn't necessary. Therefore, learn to put aside personal displeasure with your boss and focus instead on continually striving to enhance the relationship, rather than being an active party in tearing it down.

Never do the rules of engagement suggest the sacrifice of personal dignity. A little pride, perhaps, but certainly not one's dignity. While on the subject, let me address pride as it pertains to this particular issue.

Some managers have a tendency to mistake stubborn pride for professional conviction. First, they seriously dislike their boss. Period. So, in their mind, they consistently question anything the boss may decide and feel some distinct conviction to avoid as much cooperation as possible. Most often this indirect challenge is a matter of pressing the envelope to its limits. Managers will do what they are told to do, but only to be viewed as compliant. They will never do more, even if it betters the operation. They will take every opportunity to criticize the boss' decisions to others, and they will do everything they can to make the boss look as bad as possible.

This type of relationship can impede successfully taking a manufacturing firm from a conventional batch manufacturing mode to a level of waste-free manufacturing. Therefore, while focusing on the waste-free and/or lean manufacturing principles, leaders must focus on establishing and maintaining a workable relationship with their superiors. This will provide them adequate support from their boss (or bosses). The following is the most important advice I can give in assuring this happens:

Understand and actively support the number one and two key objectives of your boss and the company's primary taskmasters. The primary taskmasters include people like the President, CEO, and others in high places. Your boss' objectives will usually be easy to understand. In fact, your boss will normally make these clear. The two objectives of your company's taskmasters, however, are not always as clear but will likely be set by two to three top people in the organization. These key people determine and control the company's mission and, subsequently, the key initiatives the company will (or will not) undertake in the pursuit of its mission.

If your boss and primary taskmasters see you as supportive of the mission and their initiatives, they will support your objectives as long as these do not conflict with theirs and what they view as critical to the success of the operation.

To stay focused on this matter, understand and support the two top objectives of your boss and primary taskmasters. Though this may seem

simple enough, I can assure you it requires careful thought and consideration. A good balance between the top objectives (your boss', the
taskmasters', and yours) must be constantly maintained.

Some would tell you that you do not have enough time and energy
to focus appropriate effort effectively on all of these simultaneously.
I disagree. Allow me to share a story I heard to explain my position.

A time management expert was conducting a seminar with a group
of executives. Near the conclusion of the seminar, he pulled out a glass
jar and proceeded to fill it with some reasonably large stones. When
finished, he held it high in the air so everyone could see and then
asked: "Is this jar full?" Everyone agreed it was.

"Really?" he asked. After pausing to allow them to reflect, he reached
down, produced a handful of pebbles, and poured them into the jar.
These settled nicely into the spaces between the larger stones. He held
the jar in the air and asked again: "Is the jar now full?"

This time around most of them weren't so quick to concur. The
instructor gathered a handful of sand and proceeded to pour it into the
jar between the stones and pebbles. He asked them one last time: "Is
this jar now full?" No one was brave enough to respond with anything
more concrete than "Well, perhaps."

The rocks should represent your primary objectives. This still leaves
space for the sand and the pebbles (representing the things others
want them to give time and energy to) and even then it's hard to be
certain something else can't be squeezed in.

You should understand the top two objectives of your boss and
your primary taskmasters, and you should reflect on these at least
once each workday. Then ask yourself if you have done anything that
has constructively helped to advance those objectives. If not, then get
to work on it. You can and will find space for this in your objectives
jar if you want.

Having discussed what I believe is important to insure a good working relationship with your boss, I want to mention one last thing. Nothing written here implies you should blindly and quietly follow an
initiative your boss has set forth if it is not good for the business. In
fact, you should do just the opposite. You should resist. But you should
do this professionally, using all tools within your power to influence a
change in thinking and, subsequently, a course correction. Sometimes

this means standing tall in the face of serious criticism, potentially from your boss and others. But if the business is headed down the wrong path, and you know it, you have an obligation to make a difference. Of course, if all else fails, you have the option of going over your boss' head. You should only do this as a last resort and for something with grave consequences for the operation. While odds are you will not be placed in this kind of terrible position, it could happen at some point in you career. If it ever does, this is where the rubber meets the road and will prove how conscientious you are.

> For a lingering number of weeks after the incident, Jim found it difficult to focus on anything else. It and the subsequent series of events seemed like a nightmare that wouldn't go away. Now, three months later, it was starting to fade in memory, and fewer people were making the effort to mention the subject.
>
> Overall, business conditions were good. There were no significant delivery or quality issues, a contract with the union had been negotiated, and Denning—thanks to Jim—had now reclaimed a piece of Aklin, with a possibility of increasing the business again.
>
> On the achievement side, however, Jim was unsatisfied. In fact, he was feeling somewhat like a failure. He had now been at the helm of the operation for almost two years and, in truth, he'd seen few changes he had hoped for come to fruition. In thinking it over, he realized that he hadn't expended enough time and effort to bring about those initiatives he had so desired. First and foremost of these was to lead the operation out of an old, obsolete system of manufacturing through the insertion of sound lean manufacturing practices.
>
> His previous job at Crafton, under the direction of Frank Zimmer seemed so distant. He remembered the countless improvements made after Zimmer introduced what he called the Crafton Production System. It came about after Zimmer had hired an expensive group of retired Japanese professionals who were teaching the principles and techniques of lean manufacturing.
>
> While it was on his mind, Jim took the time to scuffle through his phone and business card files until he found what he was looking for. The old business card read: Marihiama Consulting Firm. On the back of the card was a scribbled Jeff Simpson and a phone number. Jeff was one of the young American consultants hired by the Marihiama group to serve as liaison with American-based organizations, which had a hard time understanding the consultants' lessons even though their English was proficient.

Jim had remembered how shocked most of the participants at Crafton were with the consultants. The Japanese, who were without a doubt experts in their field, had a brass, direct approach, which irritated and embarrassed as much as taught and converted. Some of the outspoken hourly participants had not so jokingly commented they were convinced this approach was some bitter holdover from World War II and the bombing of Hiroshima. Before long, everyone understood the consultants' disciplined approach came from a work ethic with a serious and focused approach for a significant change effort. However, people often failed to understand that right or wrong, shock therapy was part of their plan. The intent was to convince the audience change wasn't just needed but was essential to the future of the operation. Although it was a controversial technique, it had worked.

Reflecting on why he had been so lax in getting this important initiative underway, Jim listed many reasons, such as the daily distractions that restricted his ability to accelerate the things he really wanted to achieve. On the other hand, he thought, *Who's fooling who? If it had been a priority and something I wanted to happen, it would have. But how could I have foreseen losing a major customer or the problems getting a union contract or antici-pated the difficulties I experienced as a result of the accident and the following circus of events?*

Jim proceeded with the mental sparring before he convinced himself he had to get back on track and do it quickly. If not careful, he was going to wake up one day and wonder how he had worked so hard and gone through so much, but achieved so little. The truth was he was beginning to see how many plant managers get sucked into the mire of complacency. It was easy to do. *You can almost become lulled into it*, he thought.

On the other hand, Jim had a special problem. No one at the top had a dri-ving ambition to see that lean manufacturing was incorporated at the Den-ning manufacturing level. Too many other initiatives were drawing time, energy, and effort away from this important objective, and as long as this was the case, Jim knew he had some formidable resistance. In order to make it happen, he had to get the attention and the full support of at least one highly ranked person at Denning.

The obvious person to address was Patricia, but she was not going to be an easy sell. According to her priorities, DQS2000 held the top ranking.

He knew he could go over her head to the top, but that would be a dan-gerous move and one he didn't cherish pursuing. Somehow, he knew he had to get her attention and set out to do just that.

He asked Linda to call and find out if Jeff Simpson was still with Mari-hiama. If so, he would like to speak with him. Linda called and told Jim she discovered Jeff had left the company and started his own consulting firm, but another Marihiama representative would be happy to return his call. Linda said she told the lady she would get back to her, not wanting to commit to anything definite.

Jim decided to have Linda call the representative back, and then went about business as usual. A call came late that same afternoon from Hank Beckinridge.

"Mr. Warring? How could we be of help to you?"

"I was with Crafton Industries a few years back when my boss at the time, Frank Zimmer, hired your consulting firm to conduct some training."

"Yes," Beckinridge interrupted to make a point. "We still follow up with them occasionally."

"I see," Jim said. "I assume Mr. Marihiama and the others who were doing the consulting at that time are still with the organization."

"Oh yes. They certainly are. But the business has grown substantially, so Mr. Marihiama and four others with the original group spend their time exclusively on executive insight and counseling. I am one of 12 others who now do the standard, formatted training for the Marihiama group," replied Beckinridge.

"Executive insight and counseling?"

"Yes. That particular work is geared toward interfacing with top executives, with the intent of bringing them up to speed on the necessity of a world-class approach to manufacturing. In other words, helping them understand and realize the importance of shifting their manufacturing operations from a batch mentality to one piece. Along with that, we provide the opportunity for the other substantial benefits of the Toyota Production System to be incorporated and improve overall performance."

"Well, I assume that has been effective for them?" quizzed Jim.

"Absolutely. The key, of course, is for them to have the opportunity to spend some quality time with top management. Once that's accomplished, it's usually a done deal."

"And why is it, I wonder, that it often takes something like that to get management's attention?" Jim proceeded.

"Well that's a good question, isn't it?" Beckinridge responded. "I suppose everyone is guilty of having to hear the facts from others, outside their own company, before they'll accept it as truth."

"It's a shame, but you're right," replied Jim. "Mr. Beckinridge, do you happen to remember a Jeff Simpson who used to work for the Marihiama group?"

"As a matter of fact, I do," replied Beckinridge.

"What can you tell me about what he's doing now?"

"About all I can tell you is he left us about a year ago to form his own firm."

"I see. So, he's a direct competitor of yours?" replied Jim.

"That's correct," he replied without further comment.

"Do you happen to have the name of the company he's started?" Jim proceeded, realizing he was pressing the issue a bit.

"Afraid I can't help you there," Beckinridge replied.

Jim thought *What a liar.* However, Beckinridge wasn't comfortable with the conversation, Jim left it at that.

"Mr. Beckinridge, I appreciate your help and we may be getting back to you," Jim stated.

"I would be more than happy to drop by your operation and meet with you and anyone else you would like. If nothing else I could send you some literature."

"Thanks, but that won't be necessary yet. Perhaps at a later date," Jim interjected politely.

Jim was convinced he needed a *mover and shaker* to help get management's attention. However, he wasn't sure how to go about it. Certainly the Marihiama Group was respected and Owens—possibly even Brooks—had probably heard of them. If he could get them in the front door, so to speak, and into a formal setting with Brooks and Owens, success was likely.

He knew the proper protocol would be to approach Owens, get her agreement to meet with whomever he decided to recommend (be it Marihiama or someone else), and then hope for the best. He knew this approach had little chance of succeeding. As much as he liked Patricia, she had her own agenda, which didn't necessarily agree with what Jim felt was best or important for the operation's manufacturing branch. He was convinced if it were to work, he would have to have someone who could get to Brooks, in a setting where he would be willing to listen. At the moment, the prospects of achieving this didn't appear bright.

But assuming he somehow managed to set the stage in making this happen, he realized he could be treading on dangerous territory. First, he would be going around his boss, which he found distasteful. Second, he had no guarantee this tactic wouldn't backfire at Brooks' level, thus leaving Jim in more of a predicament.

Before deciding anything, he wanted to speak with Jeff Simpson. He had come to like Jeff when they had worked together at Crafton, under Frank Zimmer's direction. Jim realized that if anyone knew of Jeff's whereabouts, Frank would be the man. He had Linda call Frank to set up a time to chat.

"Well, Mr. Warring, just how has the world been treating you?" Frank began with friendship and respect.

"Absolutely fantastic, Mr. Zimmer," Jim replied. "And you, sir?"

Before getting down to business, they paused long enough to enjoy a hearty laugh with their formal introductions and then followed with questions about how each other's families were doing and the like. Jim proceeded with the business at hand. "Frank, you remember Jeff Simpson don't you?" Jim asked.

"Of course. He formed his own consulting firm about a year ago, by the way."

"Yes, I heard. That's why I'm calling. Do you have his company's name or a phone number where I could reach him?"

"I can do you one better than that. He's here in the plant today," Frank said.

"You're kidding," replied Jim, who couldn't have been more surprised.

"I'll have someone bring him to the phone now or have him give you a call later. Whatever you like, but I'm sure he would be delighted to speak to you again."

"That would be great, but let me ask something before you do that," requested Jim.

"Sure."

"Have you stopped using Marihiama completely?"

"The truth is I did stop using them. They're too expensive and, on top of that, too many other good consulting firms are out there that know the subject matter. Jeff, of course, was one of the better in my judgment."

"That's good to hear. What is Jeff doing for you? You must be in the advanced stages of the process by now."

"Don't be so certain about that, Jim. If anyone's in the advanced stages, it's Toyota. They've been at it for over 40 years and they're still improving. We've made some decent progress, but we still have a long way to go."

"That's an interesting observation, Frank. But if you have a long way to go, then believe me, we haven't even started on this end," replied Jim.

"That's unfortunate, my friend. What do you plan to do about it?" Jim knew Frank well enough to know he wasn't kidding. In fact, Frank was probably one of the more serious people Jim had ever known when it came to the aggressive application of lean manufacturing principles. In fact, Frank was probably as close as one could come to being a bona fide fanatic on the subject.

More than once, he had told Jim—with reference to the increasing number of U.S. manufacturing bases moving to other countries—that a rude awakening was going to occur. He was convinced the so-called *new economy* wasn't going to support the same solid job security that America's manufacturing had chiefly been responsible for providing.

"Frank, I have to tell you I don't have any firm answers now, other than having made up my mind to do all I can to get our top management's attention and push the process forward as energetically as possible."

"My god, Jim, you sound like a Washington politician. You'll have to excuse me for saying so but you're talking in circles."

Jim was slightly offended by the stinging comment, but remembered how Frank had a direct and sometimes chaffing way of airing his thoughts. He had always been fast with conclusions, which often were on target. Through it all, Jim and others who knew Frank highly respected his opinions and thus gave him the benefit of the doubt when it came to personal criticism.

"I'll give you this, Frank, you haven't changed a bit. You're still not one to beat around the bush. The truth is I've had a solid plan all along. The problem is that I'm fighting city hall. More new initiatives are being seeded here on a monthly basis than I ever remember seeing at Crafton," replied Jim.

"And why do you suppose that's the case? Do you really think Crafton is any more immune to new programs and initiatives than Denning or any other company?"

"Maybe not. But it sure seemed that way," answered Jim.

"Well I can tell you it isn't. The fact is, you probably wouldn't believe the number of times I've had to put my job solidly on the line to insure we kept things moving on the right track around here."

"I guess I wouldn't," admitted Jim. "So, what's the formula for knowing when to give a little and when to stand firm?" asked Jim.

"That's a tough one to answer, but I'll share something with you that could be helpful. When I started out in manufacturing, a man I came to respect highly, who was at the time a department manager and a peer of my boss became my sort of self-appointed mentor. He once said something I've never forgotten and that was over 30 years ago. He said, 'To be successful in any job, you have to understand what your role should accomplish and then make certain you never allow your fears or aspirations to get in the way of doing the right thing.' Over the years I have applied that advice. If you think about it, there's some real food for thought in the statement."

"So, don't keep me in suspense," Jim responded.

"It's never others who stand in the way of getting something important done. It's your own fears or, in the worst case, your aspirations. The fear aspect has to do with the possibility of losing favor with those who have a strong influence on your career. The matter of aspirations has to do with those things that can make a coward of you. This happens when your position, or more likely one you want, becomes so important that it distracts you from pursuing what you instinctively know is important to the job."

"That's an interesting observation," responded Jim.

"Just think about it. All right?" replied Frank.

"Yes. I will."

Soon, they had "run out of spit," as Frank used to say and then Jeff came to the phone. After a warm welcome and talking about old times, Jim got to the point

"Jeff, I need heavy firepower from the outside to help me get executive management's attention. If I could set it up, would you be interested in visiting, touring the facility, and then spending some time with our senior management to share your observations?"

"Sure," Jeff responded. "I'll be wrapping up a project here at Crafton next week and I don't have anything scheduled for the following week. After that I'm headed off to Europe for two months. So, if we're going to do something like that, it needs to be during that week."

"That might be too soon to get something arranged, but I'll get back to you on it."

Jim concluded that once again, Frank had been right on target. During the conversation, Frank had sensed Jim was struggling with what to do and wanted to make certain that he offered some good advice and counsel. And since he knew that was why Jim had called in the first place, Frank was more upfront about giving his opinion.

Jim felt like kicking himself. In essence, he'd been so wrapped up in the mire of keeping the business going and pleasing as many others in the process, that he failed to follow through on what he knew was most important to his position. He was convinced the company was headed down the wrong path and was going to continue to watch its market share erode unless he could somehow convince management to implement drastic measures needed at a manufacturing level.

When he raised the matter with Patricia, she seemed reasonably supportive. As a result of telling her he had already begun a new production approach, begun before Patricia was to push DQS9000, she suggested both initiatives could be achieved within what Jim would call the Denning Production System. However, her actual support of the process was, at best, tentative and at its worse, nonexistent. What he had hoped to achieve, when he had hired Rex Anderson and moved Joe Thompson, was far from complete. Admittedly, the plant accident and his subsequent injury had taken a toll on his focus, but this couldn't excuse progress.

He had to admit that the lack of accomplishment in lean manufacturing was a flaw in his career and one he had the personal responsibility to correct. The next day he set out to do just that. However, he would discover he didn't have many avid supporters, and that he was, in fact, entering a crucial stage where he would face some of the hardest choices of his career.

WARRING SCORECARD, PART 5

As Plant Manager

Overall, Jim handled himself well. He responded effectively in bringing some of the Aklin business back and, thus, set the stage for the possibility of regaining their entire business. He did this even with an unfavorable initial response from senior management and marketing. In addition, he devised an idea and followed it by assisting in settling a growing labor dispute, through a win-win proposal for both parties.

As things can and often do happen, the plant experienced an unfortunate accident, which could have taken the lives of some employees, not to mention Jim himself. This set the stage for a period of time where Jim, as plant manager, was both physically and mentally distracted from the mission he hoped to accomplish. Although, Jim could have done little to avoid it, and although he handled himself well through it all, he expended much time and energy that could have otherwise been directed at making more progress elsewhere.

As Chief Conductor for Lean Engineering

Even though he knew the process prior to joining Denning, Jim has brought little to the party so far. Had he focused on this more at the offset, he could have made the situation different and more positive overall. However, as is the case with many young plant managers and especially those taking on the role for the first time, the depth and scope of responsibility surrounding the job can sometimes be overwhelming.

On the other hand, Jim has, at last, recognized that the lean manufacturing he has wanted to bring to Denning hasn't progressed as he thought and that he has to react accordingly. In doing this, he sought the advice and counsel of his previous boss and mentor whom he respected and who had experience in putting the practice in place. He followed up on this by contacting the type of individual who could sell the process to the higher levels in Denning. Jim had also opened the door by approaching his superior, Owens, with the idea. Although her response was not totally supportive, Jim decided it was an initiative he was now going to pursue with all the energy he could muster.

7

Successfully Guiding
the Lean Initiative

IN JEFFREY K. LIKER'S BOOK *Becoming Lean,* (1998, Productivity Press), he states: "Everybody is doing it—becoming Lean, that is. Or are they just talking about doing it?" (Page 3) Mr. Liker continues: "For some reason, which I do not claim to understand, it took us at least a decade from the early 1980s to the early 1990s to figure out that there was more to Japanese manufacturing than individual techniques like quality circles, statistical process control (SPC), and preventative maintenance. Probably we were in denial and saw what we wanted to see." (Page 5)

And in quoting John Shook (from "Bringing the Toyota Production System to the United States: A Personal Perspective"), Liker remarks: "John Shook argues persuasively that TPS is not a collection of techniques, but a way of thinking—a paradigm." (Page 27)

In *Fast-Track to Waste Free Manufacturing,* I pointed out that too many companies in the United States talked more about becoming Lean than practicing it. I agree that too many managers and others spent too much time viewing the Lean Process as some sort of SPC or quality circle exercise. Without question, I concur with Mr. Shook's assessment that TPS is a way of thinking. Unfortunately, it's a paradigm that few American manufacturers have seriously adopted.

So, what does this have to do with guiding the lean initiative? To answer that, my intention is to set the stage for one of the more important elements of this work. I used the word "guiding" rather than leading because I believe the obligation of the plant manager must go further than just assuring the basic tools, techniques, and concepts of lean manufacturing are incorporated on the shop floor. As I've mentioned, plant managers must be the ambassadors for the process, and they must do all within their power to help build it into every facet of the business, from order to delivery.

We must realize that for every success with lean manufacturing, there are just as many (and perhaps more) total failures. Much of this comes from companies' impatience to not recognize and accept that Toyota had to dedicate decades of unerring commitment to achieve a true competitive edge. Therefore, it's important to examine what to do when you find yourself in such a situation and the choices you have. To do this, we will start with handling the "ghosts from the past."

DEALING WITH GHOSTS FROM THE PAST (THE SHADOW INITIATIVES)

There comes a time when all plant managers face difficult career decisions. In almost every case, they must come to terms with those decisions on their own. Someone is not usually around to hold their hand and guide them through and it is when the old adage "It's lonely at the top" can become a stark reality.

One of the principal reasons for this is a business direction change and not always change for the better. Business, in general—and big business, in particular—has a way of changing priorities on a somewhat regular basis. This usually comes from new leadership, but not always. As a result, various key objectives that people within the organization were specifically hired to accomplish can become what I term Shadow Initiatives.

Shadow Initiatives are those key business strategies that were communicated to the workforce as being crucial to the company's success, but which (for one reason or another) have since been tabled, sidelined, or canceled. These are seldom, if ever, killed off entirely because someone would have to admit poor judgment and/or a lack of foresight.

What typically happens is that cursory support continues and the initiative raises its head occasionally. In the administration of company goals, objectives, and stated obligations, the initiative can cast a shadow along the way. Therefore, it's never fully forgotten but usually has a dim substance resembling a ghost from the past.

Take a moment and look back over the course of your career. Do you recall a Shadow Initiative? In all likelihood, you do because every company has experienced them. In fact, some organizations have so many and thus are so haunted with ghosts from the past that employees have no confidence the company will complete anything it starts. Think how difficult it then becomes to convince those same employees it is critical for them to take up arms and actively support what's been outlined as the latest cure-all.

When you repeatedly hear the rank and file complain about management's lack of commitment and/or direction, you generally have a situation where the company has started numerous initiatives and dropped them quietly in pursuit of the newest "improvement fad." These ever-changing priorities are caused by poor business performance (e.g., missing profit projections or losing share of market.) However, they can just as easily come during better times and because of someone who has top management's ear. Somehow, management becomes "sold" on pursuing a new initiative (often broad in scope) which has the need for active employee participation. This, in turn, can have a negative impact on previous pursuits that were once outlined as crucial.

In Chapter One's "Clearly Understanding the Keys for Success," I emphasized the importance of commitment and focus on the part of plant managers. As this applies to an undertaking that restricts the prime objective they want to accomplish, they can be faced with a career decision that could change their entire career. It is with regard to this that I have some thoughts.

First and foremost, make certain the new undertaking doesn't go against the grain of what you were brought aboard to accomplish. The newer initiative could be a needed refinement that fits well with your directive and design. You must ensure you do not allow pride of ownership to become a stumbling block to good judgment. On the other hand, if it is in opposition to your key objectives you are essentially left with three choices:

1. *Stay and compromise.* Even though this may seem the easiest choice, it isn't always the proper one. If the change in direction is a needed refinement, then swallow your pride and go along for the ride, so to speak. On the other hand, if it isn't and you comply

solely out of fear of how you will be perceived by the powers above, then you've made the easy choice but what could well be the gutless one. If this happens, I assure you those who report to you will notice.

2. *Stay and fight.* Assuming that without question the change is going to go against the grain of your professional values, then this is your option. Understand, however, if you make this decision, you may or may not win the battle and you could lose the war (your job.)

3. *Move on to greener pastures.* Though this decision has its time and place, you must take care not to make a career of running every time a serious challenge develops. When done repeatedly, this tactic can taint a professional reputation.

Circumstances will vary; but as you can see there are no miracle solutions. Therefore, it most often boils down to the professional risks you are willing to take, in order to see the things you hold a strong conviction for are carried through. As I've said before, a plant manager's position is indeed a sizable challenge, wrought with difficulties and uncertainties. On the other hand, in terms of a feeling of accomplishment, it can be one of the more rewarding.

MOVING FORWARD AGGRESSIVELY WITHOUT FORMAL APPROVALS

I'm going to make a mind-boggling statement, but bear with me because it is not that far-fetched:

Most often, the lean manufacturing effort that reaps the biggest success is the one where someone in charge assumes the responsibility to aggressively implement the process, without formal approvals and before the wheels of bureaucracy set in and slow the process to a stall—if not ultimately a slow death!

I'm in no way suggesting plant managers should do what they like, when they like, and everything else be damned. However, in reality, no plant manager has absolute authority to use the tools and techniques outlined in Toyota Production System (TPS) as they see necessary. That would be, of course, unless a position existed where every decision had to be cleared by someone else, down to the nth degree, which isn't likely. What I'm suggesting here is moving forward aggressively within the recognized, prescribed limits of a manager's power and authority. Nothing more.

When the issue becomes one of formal approval before proceeding (which admittedly is required before undertaking many processes and for those that deal with new or revised products, healthy expenditures, etc.), you are setting yourself up for trouble. This is because the bureaucracy can slow the process dramatically. Even worse, the process could wither and die on the business vine.

I'm proposing that to lead the lean initiative you have to proceed as the Nike commercial said and: "Just do it!" However, it is also important to do it aggressively at the offset. If you do, you are going to make substantial improvements that few would argue should be tabled, restricted, or canceled. Further, doing it before the wheels of bureaucracy set in will swing considerable influence your way because no company, with any good business logic, will require you to stop an activity proven to be of substantial benefit to quality, productivity, and customer satisfaction.

The key is to know the importance of becoming lean. From there it's a matter of communicating the concept appropriately, organizing support for the effort, and getting the job done quickly!

I can assure you that any plant manager who does not fully understand the importance of lean manufacturing and does not place this at the top of his or her arsenal will not be successful. However, just as important as recognizing the need is being aggressive in its pursuit. When it comes to lean engineering, the speed of implementation is paramount because the manufacturing sector of our economy will continue to decline without it.

To be totally realistic, however, we must consider a scenario where the plant manager recognizes the need but doesn't have the knowledge and/or experience to be aggressive. What to do? The answer can be complicated because, after almost 30 years since TPS's introduction to U.S. manufacturing, a person placed in charge with no lean engineering expertise can be a severe handicap. The only choice is to get a quick education (via seminars, etc.) and to hire and staff the talent required to support the effort. Though another alternative would be to hire outside consulting expertise and turn the job over to this group, it has been my experience this isn't an effective approach.

Today, if a company knowingly hires someone to take over a manufacturing operation without solid experience in lean manufacturing, it is taking a risk. In fact, if I were the head of a company with manufacturing operations or divisions, I would set firm policy requiring that all new plant managers be proficient in lean engineering. I would go further to insure that any existing plant manager, who didn't meet those specifications, be required to gain the appropriate knowledge quickly through company-sponsored education and/or training.

SOME IMPORTANT OBSERVATIONS BASED ON PRIOR LEARNING

During the latter part of my career with United Technologies, I taught and assisted others in how to implement the basics of lean manufacturing. I thoroughly appreciated being given that opportunity because along with it being a rewarding and educational experience, it allowed me to see the "real world" when it came to the more common obstacles surrounding this effort.

Though I had educated myself in lean manufacturing concepts, principles, and techniques prior to joining this effort, it was during this experience (while working around the world in a wide variety of manufacturing) that I received a "masters degree" in the process.

My fellow team members (Barry O'Nell, Ed Cannon, and Gary Roscoe) and I saw everything from fear of the unknown to serious concerns by employees about job security. We encountered plant managers who actively and energetically supported the process and those who displayed everything from passive support to complete resistance.

Through it all, the process prevailed. The point is, given a reasonable amount of top-down support, if a company can maintain focus on actively pursuing the tools and techniques of lean engineering, it will always make significant improvement no matter how much resistance is encountered. The process always works, regardless of the industry type, how long it has been around or how set the employees may be in their ways.

I learned much during this period in my career related to the overall process of implementing lean manufacturing and I would like to share some observations and a few helpful hints:

General employee attitudes toward change: There is little to no difference in what shop floor workers (most often called production associates) hold as convictions about their job and a willingness to change. Frankly, I had a hard time seeing major differences in the average worker in Europe, South America, the Far East, or the United States. While much has been written and discussed on this topic—especially in terms of the Japanese mystique when it comes to "them versus us"—I believe this has been overplayed. Simple reasoning would indicate that cultural differences exist, but my experience is when significant attitude problems surface, you will most likely find they come from the middle management ranks rather than production associates.

> *Helpful hint #1*: Focus your initial efforts on selling the lean manufacturing process to middle management and keep a sharp level of attention on what it takes to gain its active help and support. In most cases, this will call for special training and/or communication sessions, specifically designed for the middle manager.

Understanding the real drivers behind employee resistance: If a company is having serious problems with employee resistance, the culprit usually isn't radical employees. Employees fully expect management to prescribe and determine how to accomplish the process of production and to stay abreast of the latest and most effective ways of achieving this. When a potential problem with the process surfaces affecting long-term job security, then major concerns and/or resistance can surface. If you find strong resistance for any other reason, you can usually chalk it up s a major shortcoming in communications.

> *Helpful hint #2*: Whenever possible, clearly communicate to the workforce that the improvement process does not have a primary mission to eliminate jobs. While manpower adjustments may be required, and often are, many organizations find ways to use displaced workers in new and important assignments that improve overall productivity. If a plant has an over-staffing problem, then handle and communicate it as just that—a needed reduction of resources. Do not damage the sanctity of the Lean process by making it the enemy of the worker. You cannot ask production associates for their help in

implementing a process, then use it as a means to eliminate them. Such an approach is doomed to fail and could backfire.

Breaking down old paradigms: Initially making a production process change, under the direction of a corporate-sponsored training session, is easier than making lasting change. Lasting change only comes by ridding the workforce of old paradigms nurtured under a different production system. The most successful operations are the ones with strong local management commitment focused on maintaining and improving the process.

> *Helpful hint #3:* When beginning communications and training, ensure production associates, first-line supervisors, unit managers, et al., understand that one of the more important lean process objectives, where they are key players, is to insure the basic disciplines are consistently maintained.

Structuring the process to fit key operational needs and requirements: Local management must tailor the improvement process to fit a particular emphasis structured to provide the greatest support from employees and to provide the fastest favorable impact on customer satisfaction. This, of course, must be done without sacrificing any of the stated principles of the improvement process. For example, a given operation might have a good reason to focus on "mistake-proofing" (poka-yoke) initially because of a history of continuing problems seriously disappointing customers and, therefore, declining market share. In doing this, the operation may decide to table immediate concentration on the implementation of one or more Lean tools (e.g., Kanban). Conversely, another operation may have a different problem/opportunity and, therefore, would have a different outlook on the required initial focus. Remember, nothing is sacred about what tool to use first or when to incorporate it. However, it is also good to remember to keep the overall process "holy" and at no time should a decision be made to entirely drop, change, or amend any of the established principles.

> *Special note:* Four driving principles exist in waste-free manufacturing (as opposed to three in the TPS): Workplace Organization, Uninterrupted Flow, Insignificant Change-Over, and Error-Free Processing.

Though there is some disagreement among specialists in the field regarding what the principles of TPS precisely are, they most often define them as: Takt Time, One-Piece Flow, and Pull Production.

THE WARRING ADVENTURE, PART 6

Jim had gone so far as to approach Owens about bringing Jeff Simpson in for his observations and ultimately a discussion with her regarding his findings with what he might propose as a means of further improvement. Owens rebuked him, and for the first time he could remember, she had lost her temper. She accused him of not seriously supporting the DQS2000 initiative, which she was convinced was the answer to everything. Further, she remarked, she wasn't interested in hiring a high-priced consultant: "He would only take precious time picking everyone's brain for ideas and then sell them back to us as his," she insisted.

In the end, she agreed Jim was welcome to bring him in to look at manufacturing and to give him some ideas. That was, of course, as long as the man was willing to do it as an exploratory venture. Again, she emphasized she had no desire to talk to Jeff.

Not wanting to turn the opportunity down, Jim did have Jeff in for one day. At breakfast, before visiting the plant, Jeff pointed out that he and his firm specialized in carrying the lean engineering process far past the manufacturing shop floor. They were working at driving it into the total business chain of events, from order to delivery. He further pointed out this was what he was working on at the time at Crafton Industries, for their mutual friend, Frank Zimmer.

After Jeff saw Jim's operation and spent some time with Jim's staff, he went over a checklist of his observations with Jim and gave him some literature on his consulting firm. He had three conclusions. First, that Jim, after more than two years on the job, had not taken lean manufacturing as far as he should have. Second, even if he had, no mechanism was in place to carry the benefits into the total business chain and to the customer level. Finally, he concluded DQS2000 did not go far enough. Though he agreed it was a good quality assurance initiative, Jeff called it a "lightweight" business strategy.

After Jeff departed, with a sincere thanks and a promise from Jim that he would be hearing from him, Jim proceeded to put together a special report which he forwarded to Owens. It included Jeff's observations and Jim's take on what he had to say, along with the literature Jeff had given him.

A month later, Jim was still waiting to hear from Patricia. He began to think that he should have been more selective with what portions of Jeff's observations he included in his report. He was certain Owens wouldn't appreciate Jeff's insight into DQS2000, and she probably thought Jeff only wanted to get his foot in the door and extract money for an extensive consulting venture.

In the meantime, DQS2000 was proceeding quickly throughout the entire business enterprise, at great expense, and as an ever-growing requirement on the time and energy of almost everyone.

A short time later, a cold winter-like spell was cast inside the walls of Denning Corporation. It was to become a day of serious reckoning for all concerned.

Late spring was when Denning was in full swing and at its highest annual employment level. Seasonality had always been a factor in the business, with most of Denning's customers gearing up to maximum requirements in the May-June time frame. As a result, the manufacturing end of the business, under Jim, had recently gone through a major recall of production manpower.

Jim got the bad news in a frantic call he received from Nelson Ambey, the chief manufacturing customer coordinator. Patricia Owens and her staff had set up the position under the umbrella of the new DQS2000 initiative.

Nelson proceeded to tell Jim that he had just been informed by Aklin it was reducing its agreed upon product to a minimum, effective immediately. This came just after Jim had received a formal letter from SJM Industries, another important customer, stating it was reducing requirements and strongly hinted it was planning to cease business with Denning at the end of the year.

In both cases, delivery problems were the culprit. Under Owens' direction, the basic production scheduling philosophy had been shifted from "build to order" to "build to forecast," based on historical projections of annual requirements. In the process, Denning set up four new regional warehousing operations around the country and packed them with inventory. Unfortunately, they didn't have the particular product mix most customers needed. In fact, during the six months the new scheduling scheme had been in effect, Jim watched customer complaints steadily rise. Even though he had actively questioned the new strategy and had, in return, been criticized by Owens for not having an open mind, Denning clearly had a serious problem.

With the news at hand, Owens scheduled an urgent meeting with her entire staff, and Jim sat waiting for the proceedings to begin. There were looks of concern on the faces of some in the room though most went about extending the usual lighthearted gestures and jokes common before such a meeting.

Owens began the session with a rambling report on the status of the Aklin and SJM accounts. She speculated why the sudden downturn in requirements, punctuated throughout with notations geared at insuring Denning's actions did not receive the blame. She had even accused Aklin and SJM of ineptness for not gearing their requirements to what Denning had "gone out on the limb to produce and warehouse for the customer's convenience."

After opening the meeting for discussion, Jim sat patiently as some of those quick to speak gave their insight and speculation, all of which seemed to agree with Owens completely. One other person in attendance, besides Jim, felt differently: Fred Johnson, the Sales and Marketing manager. He was a peer rather than a direct report to Owens. Jim reflected, as Johnson spoke up, that Fred had been one of the first managers he had spent some time with when he arrived at Denning. He had come to like and respect him.

"Well I, for one, don't think we're on the right track," Johnson began. "For us to sit here and blame the customer for not using the products we think they should and for any subsequent dissatisfaction they may have with us is way off the mark. In fact, it's ludicrous."

"We're not saying that," Owens retorted. "We're just pointing out that history doesn't lie. We spoke with them regarding the projections on the front end and they basically agreed on what we produced and stored."

"Of course, they agreed to it. It was all they knew to do at the time. But they didn't sign a contract on it and they didn't tell us to build product before it was needed and then warehouse it. That was purely our decision, not theirs. Let's be frank. We goofed on this one."

"I have to agree with Fred," Jim interjected.

He was sitting at the end of the conference table where everyone was gathered, with Patricia seated at the opposite end and Fred just to her left. Everyone had been more or less facing toward the front, focusing on the conversation between Owens and Johnson when Jim spoke. They all turned their heads toward Jim and remained silent, as if waiting for a further explanation, since any challenge of Owens' convictions was out of the ordinary. Patricia gave him a hard stare but, like the rest, waited for further comment.

"Regardless of what we think caused the problem, I'd have to respectfully say that all the finger pointing isn't going to resolve that our production requirements have changed and we're going to have to decide what to do with all the new people we've brought aboard recently. And, if we want to talk about history, the facts are that we've always had constant change in product mix and I can testify to that, coming from manufacturing. It's been a way of life—."

"Hold it, Jim," Owens demanded, her voice rising and edgy.

"We're not pointing fingers at anyone. We're just stating the facts behind what's brought us to this point. As far as a way of life, so to speak, that's the whole idea behind what we've undertaken. To change our way of life to be more effective, for us and our customers. And you should remember we've got other issues outside just your area to deal with here," she concluded.

Fred Johnson cleared his throat quite loudly and everyone turned his way.

"I assume you have something further to say, Fred?" she quickly challenged, giving him the same hard stare she'd earlier given Jim.

Johnson looked at Patricia for a moment and then replied, "I've said it." With that, he stood, gathered his things, and left the conference room.

After an excruciating silence, Patricia took a deep breath and let it escape as a clear indication to everyone of her growing frustrations.

"Let's get on with it," she said, deciding it was best not to get into a discussion about Fred's quick departure. However, some were visibly shocked by the proceedings. Jim felt a little sorry for Patricia. She was clearly perplexed and disappointed with the events; especially coming at a time when she was just getting DQS2000 off the ground. Jim was convinced she was a good person and a dedicated, hard working manager. However, he was just as convinced she was ill-advised and, therefore, on the wrong track as to the primary focus for the business.

The meeting continued for some time, as Patricia reinforced her position, and most of the group energetically agreed. She stressed what they needed to do was "educate the customer." She was adamant when customers "learned the system, they would willingly fall in line because it was "to their best benefit."

All this only reinforced Jim's earlier observations about Owens. As the group rambled on about customer education, he was convinced he was witnessing a team so obsessed with completing a mission, they were willing to do it at all costs, up to and including losing customers in the process.

It was at this moment that Phillip Brooks unexpectedly dropped in on the meeting. He took a seat in the one recently vacated by Fred Johnson, noting he had just dropped by to see how things were going and asked that they proceed. The conversation regarding the need for customer education resumed, and Brooks listened patiently before interjecting a thought.

"I agree we have to help our customers understand that we want to serve them better. That's important. But we can't forget if we aren't satisfying their absolute needs and requirements, then we could be forced to readjust our thinking. I believe this calls for outside help, some expert advice to prove we're headed down the right path."

"I know just such a person," Jim interjected.

"Oh? And who would that be?" Owens responded.

"Jeff Simpson's firm. You remember, I sent you a report on his findings when—."

"Jim," she impolitely interrupted, "we've already been over this—"

Owens stopped as a result of Brooks raising his hand slightly in a silent but clear gesture for her to hold on for a moment.

"I think I'd like to hear a little more about this, if you don't mind. Perhaps after this meeting, you and Jim could bring me up to date," said Brooks, directing his remark in a courteous manner, considering the setting, to Owens rather than Jim.

Owens never took her eyes off Jim's as she replied she and Jim would be glad to drop by. Jim was convinced he really put his foot in his mouth, but he wasn't sorry he had spoken. Shortly after, Brooks left the meeting without further comment.

Owens wrapped things up without a game plan for addressing the business loss, but scheduled another meeting for early the next morning. She closed by asking Jim to stay so they could proceed together to see Brooks.

After everyone had departed, Patricia remained seated at one end of the conference table and Jim at the other. She took her time to stack all the papers she had in one pile before slipping them into her briefcase. Only then did she lean back, fold her hands, and look squarely at him.

"I have to tell you, Jim. You've disappointed me."

"I'm sorry, but I'm not at all sure why," he responded.

"Oh, I think you're perfectly sure."

"If I've offended you by mentioning Simpson and his firm, then I'm sorry. But I felt it was important and it certainly wasn't meant to instigate a challenge."

"It's not a matter of instigating. It's a matter of loyalty," she replied.

"Loyalty?" Jim quizzed.

"That's right, Jim. Loyalty," she replied, her forehead beginning to turn a rosy color and a frown forming on her face. Quickly standing and almost knocking her chair to the floor, she gripped the sides of the conference table with her hands, leaned forward, and proceeded. "Let me tell you something, Jim Warring. If you think you're going to attend a meeting of mine and undermine me to my peers and my boss, you have another thought coming. I have made it clear I wasn't interested in bringing your consultant friend on board—now or in the future. Apparently that wasn't good enough for you, so your response is to go over my head," she said stopping long enough to catch her breath.

Jim was starting to feel angry himself and decided he would also stand.

"Sit down!" she insisted.

That was the icing on the cake for Jim.

"When I'm ready," he stated defiantly. "I'm not going to sit here quietly while you accuse me of something like that. I've been nothing but a loyal employee. But I'm neither blind nor stupid, and you've made it clear often that you aren't willing to consider anything outside DQS. You seem to think it's the answer for everything. You can do what you like, obviously. You're the boss, but that doesn't mean I have to march in tune with it like some toy soldier. And one other thing. If your definition of loyalty is for me to act like those who are humming the tune they think you'd like to hear, then count me out."

He had no sooner said it than he was already regretting some of what he had spouted, although on the other hand it was precisely how he felt. They looked at one another. Owens straightened, turned, and started for the door. On the way out, she calmly remarked, "I think that's enough said. We better be on our way. I'm sure Brooks is anxiously awaiting to hear all about your consultant friend."

He gritted his teeth and remained in place for a moment before turning and slowly following. With her head held high, shoulders back and arms swinging in tune with the brisk steps she was taking, Owens didn't stop to see if Jim was following her.

On his way home that evening, Jim felt an equal mixture of contentment and concern. The meeting with Brooks had taken a number of sharp twists, with for a time him sitting idly by while Brooks and Owens almost squared off. However, through it all, Brooks apparently liked and respected Patricia and they tolerated a lot of give and take. Jim wished that was true for his relationship with Owens. For whatever the reason, Patricia had a totally indifferent attitude when it came to interfacing with her direct reports, or did she, he wondered?

She had exploded quickly and harshly, but when he returned the fire, she calmly proceeded. Perhaps, he thought, there's not as much to this as I believe. Although he hadn't liked her snide comment made about his "consulting friend," maybe she meant just that and nothing more. He was going to have to wait and see.

Brooks listened to what Jim had to say about Jeff and his firm. Jeff had referred to his process as an "expanded lean concept." Brooks quickly let them know that he was interested. Patricia, on the other hand, immediately challenged him, saying she didn't believe they could adequately support DQS2000 and take on another time-consuming initiative. Brooks, in turn, flatly disagreed.

Brooks argued that though DQS2000 was important to him—in fact, his baby so to speak—he more importantly wanted the correct strategy to place Denning first in the industry. Owens countered they were aggressively pursuing DQS2000 because it was geared to do just that.

Brooks shot back that if that were the case, then she had better open her eyes because it didn't seem to be working. He mentioned that two customers had already shortened requirements for the year and were talking about leaving. Owens assured him they hadn't given it enough time, to which Brooks responded they didn't have the time. She then complained about a lack of support and mentioned that Fred Johnson had stomped out of her meeting that day.

Brooks let her know that Fred had "stomped" directly to his office to inform him how upset he was with her and her "cure-all focus" on DQS2000. He went on to mention he had to smooth Fred's nerves because of the incident. One thing was apparent to Jim. Brooks wasn't going to let any initiative become so fixed and rigid that it became a stumbling block to the success of the company. He was soundly impressed with the man.

They finally agreed they would invite Jeff Simpson to meet with Brooks and his entire staff and then decide if there was any merit to pursuing the matter further. On the way out, Patricia didn't bother to smooth things with Jim, which left him wondering where their relationship stood.

So, here he was on his way home after a day he wasn't soon to forget. He had put himself on the line with his boss, in opposition to her thinking, and felt anxious about that. On the other hand, he felt great because he could tell his friend, Jeff, that he may have opened the door for him. After that, it would be up to Jeff to sell the senior management team on pursuing what was certain to be a new and sizable initiative for Denning.

WARRING SCORECARD, PART 6

As Plant Manager

One of Jim's better performances. Admittedly, some of this could have come because of more time and experience on the job. No matter the reason, he decided to take a stand on what he believed was right for the operation. He did this in spite of the pressure from his direct supervisor and fellow associates to, in essence, back off and fall in line. Buckling to this would have been the easiest thing to do. But he didn't. Further, he left little doubt he would be willing, if necessary, to put his job on the line. And when the matter was elevated to a higher level of conflict, Jim clearly displayed he was not afraid to speak his mind, regardless of the consequences.

I assure the reader I do not believe the acid test for anyone is how willing they are to engage in conflict with others, and especially their boss. But I do believe that America needs, now more than ever before, plant managers who are "energized with the need for change," as mentioned in *Fast Track to Waste-Free Manufacturing* in the "Revolutionary Pointers" at the end of Chapter 2 (page 38):

"The Plant manager's journey from Mass [production] to WFM begins with an unrelenting conviction that everything in conventional manufacturing must be questioned by repeatedly asking why and quickly eliminating anything that is not clearly value added."

I additionally stated in the "pointers" section of Chapter 1 that: "The Plant Manager is the true revolutionary—the person between top management and workers, the person responsible for changing and implementing the rules. Without his or her commitment and focus, you don't have a prayer of moving from being a conventional mass manufacturer to a waste-free manufacturer."

As Chief Conductor for Lean Engineer

There is little to comment on at this point, with the exception that the focus is finally developing in Jim. Although somewhat late out of the starting blocks, Jim is about to make significant inroads into change for the better at Denning.

8

Expanding Beyond the Manufacturing Arena

EVERYTHING I HAVE ADDRESSED thus far has dealt almost exclusively with how to change the manufacturing sector and, in doing so, putting it in tune with good globally competitive practices. This is important because manufacturing is where the improvement process has to start. However, the process must go beyond this if a company desires to reap the maximum benefits from such an initiative.

The purpose of this chapter is to show the linkage that should occur between an improved manufacturing sector and the other elements of the business. In other words, how you energize the entire business chain, from order to delivery.

I would be the first to admit that performing lean manufacturing in a few select areas of the factory would serve no logical purpose. But just as purposeless is to incorporate Lean practices throughout manufacturing and then have no plan to link this improvement to the entire business process. Unfortunately, I have seen this happen repeatedly. This is why I have done some work on what I call the waste-free enterprise (WFE).

I began focusing on this issue in the mid-1990s as a result of recognizing that the prescribed benefits of waste-free manufacturing (WFM), or any other lean engineering effort, cannot be achieved without a plan to address the entire business. This was pointed out clearly in *From Lean Production to the Lean Enterprise* (*Harvard Business Review*, 1994, by James P. Womack and Daniel T. Jones.) The article focused on the need for expanding the Lean concept throughout the entire business enterprise. I was convinced of the need and then out-lined and established the most appropriate drivers (or principles) for achieving this end—in keeping with the utilization of the Waste-Free philosophy at the manufacturing level.

However, I wanted to approach this with the same basic thrust in mind that was the catalyst for WFM—speed of implementation. I believe if the effort calls for overhauling every major business system in order to achieve a Lean enterprise, the chances of success are less than good. On the other hand, the opportunity to become more waste-free and therefore serve the customer better, is one that any committed business can achieve, without it requiring endless years to accomplish.

Much like WFM, WFE starts with a set of drivers along with a foun-dation that is the springboard for continuous improvement. In the Con-tinuous Improvement Pyramid outlined for WFM, everything is built on the foundation of the Workplace Organization. In WFE, this pyramid is constructed on the foundation of Customer Values, which as you will see is logical reasoning. See Figure 8-1.

The separate joining blocks, which serve to take the Pyramid of Achievement to higher levels, evolve around what it takes to reap a true competitive advantage. This begins with time and energy expended in implementing WFM on the manufacturing shop floor. Without these efforts, the advantages achieved in speed-to-market, quality, etc., would never get funneled into the total business chain and the overall effort would be limited in scope.

In examining the drivers of WFE, a past associate asked why I had not decided on Employee Participation as the basic foundation since, without dedicated employees who actively support the process, the chances of getting it off the ground were slim to none. Though I couldn't agree more about the need for active employee participation, employees must first understand precisely *what* they are expected to participate in achieving.

Everything must start with the stark realization that most manufac-turing-oriented companies simply are not aggressive enough in under-standing customer values and reacting accordingly. It never ceased to amaze me how far removed the thought process of the average worker in industry was from the customer. They were not at fault because it is what they were trained to do. In the past, the focus was to come to

Figure 8-1. Waste-Free Enterprise

work, check their brains at the front door, produce products to specification, and get it out the back door, as fast as possible.

The truth is, most production employees know little about how and where their products go to market and, more important, how they perform in application. Unfortunately, one of the distinct problems with manufacturing is it has little to no instantaneous customer feedback. That is not the case in every business. In our growing service sector employees can get instant feedback pertaining to customer values. A waitress or waiter, for example, will almost instantly know if a customer is highly displeased. If the server is conscientious enough to ask, he or she can find out if the customer is satisfied. In manufacturing, even though employees may see Product Field Failure Reports and other such data, they seldom get a direct response from the customer. Thus, they can only speculate about the customer's true feelings. Therefore, it becomes the company's obligation to pursue customer values and translate these into day-to-day practices and procedures.

This calls for re-engineering the daily focus (commitment) of our national workforce. This not only applies to the manufacturing sector but all areas, from order to delivery and beyond. Frankly, it would be good for the service sector to focus more strongly on customer values. I'll address my logic for that later, but for the moment, I would like to concentrate on how this issue applies to manufacturing.

Considering everything, customer values must be the foundation and the key driver, for any process aimed at achieving and maintaining a globally competitive status. In examining the basics of the WFE process, the entire set of drivers are represented graphically as individual joining blocks, or separate sections, all of which support the structure at each step along the way. This is an ideal representation because each of these must be on equal footing in terms of strength; if not, the process will become structurally unbalanced. In addition, if, for any reason, one or more of these begin to weaken and crumble, they will compromise the entire structure.

However, when it comes to business processes (the office) versus manufacturing processes (the shop floor), the chains of tradition are stronger and more pronounced. What those performing in the office areas have to understand is the grave importance of changing the way they perceive their jobs and the way in which they interface with fellow employees and customers. This calls for the company to assist them in moving from a traditional focus, which centers on the individual's performance and accomplishments, to a more ideal focus centering almost exclusively on the *team's* performance. Doing this correctly enhances continuous improvement across the entire business structure. The major differences in focus between the more traditional and world-class styles are outlined in Table 8-1.

Though most of the differences are self explanatory, a few require further insight. Under Solving Problems versus Reducing Wastes, the idea is that by working aggressively on waste reduction many of the

Table 8-1. The "Traditional" versus the "Ideal" Management Focus

Traditional	Ideal
• Performing soley one job	• Initiating teamwork
• Meeting singular objectives	• Achieving team goals
• Measuring individuals	• Tracking team progress
• Motivating people	• Removing barriers
• Rewarding individuals	• Honoring the team
• Solving problems	• Reducing waste
• Supervising employees	• Developing people

origins of recurring problems and errors can be driven out of the process. Under Motivating People versus Removing Barriers, traditional thinking places a strong focus on motivating individuals to meet or exceed their formal objectives. The new thinking is placed on removing obstacles and barriers that restrict the team from achieving more effective results. As you can see, this is a big difference in thinking.

Anyone who decides to pursue this will quickly see that the greatest obstacle to overcome in achieving this end is changing the long-standing paradigms most professionals hold about their jobs. The difficulty is multiplied substantially when the issue becomes one of changing the culture of the entire business structure.

We Americans have learned from youth to focus on individual accomplishment; so, regardless of how poorly the team performs, a given individual can receive recognition and be viewed as successful. While this may be fine for sports, it is totally ineffective as it pertains to a globally competitive business.

Employees have to come to understand that the sum of the team's performance matters. This isn't to say individual performance should not be recognized or applauded. It is to say, however, that the more highly emphasized rewards and incentives should always be for the team's accomplishments.

There is a sensitive balance in all this and it isn't something that can be approached with blinders on. Though the new emphasis must always be on the team, we have to realize that good people can work on bad teams. If this happens, these people have to be provided the opportunity to perform where they can be recognized and rewarded accordingly. Therefore, management has the never-ending challenge of accomplishing this. Given the cultural components that make up what we've been trained to believe was right and important, American-based businesses will find this a difficult task.

Having highlighted the challenges, which every company will face, I would like to discuss the WFE drivers. I will examine the drivers and their major emphasis (see Table 8-2) and then use "The Six Essential Steps" to point out how and why the principal tools apply.

Table 8-2. The Six Essential Drivers

WFE Drivers	Major Emphasis	Principal Tools
1. Customer	Values	Surveys & feedback
2. Management	Commitment	Vision & incentives
3. Employees	Participation	Training & measurement
4. Suppliers	Involvement	Policy & partnering
5. Process	Discipline	Rules & follow-up
6. Improvement	Continuous	Tools & techniques

Customer Values: Customers are mentioned first and are the basic foundation for the process because what they value starts everything. In your particular business, do you know what your customers value? Is it best price? Best quality? Overall product availability? Service? Some combination of these? All of them? Until this is clear, any strategy created to serve the customer is, at best, pure speculation.

So what's the best way of gaining such knowledge? Of course the simple answer is to take a survey. This is done frequently in industry; but no matter how many surveys are performed it is important to remember that customer values can and do change. Think about this. Before automobiles came on the scene, customers in the market for transportation placed a high value on the horse and buggy. But after the advent of the Henry Ford era and the ability to make automobiles affordable, customer values altered. This applies to new and innovative products, as well as new and innovative practices. For example, for decades U.S. auto makers made larger and shoddier cars until the Japanese came on the scene with a better approach to insuring high quality. Today, regardless of the make, model, or price of the automobile, customers expect and demand higher quality.

Seeking and understanding customer values is a never-ending process. Therefore, people who think they have gone all the way through the implementation of some type of formalized customer survey are likely fooling themselves. Being lean is a challenge of gigantic proportion, requiring an ongoing organizational focus. Doing this properly requires extending the customer communication chain far past just the Sales, Marketing, and Product Engineering functions. In some cases—depending on the business—it can include opening direct communication channels all the way to the manufacturing level and other functions, which in the past may have been held at arm's length from the customer.

WFE's whole premise is to make the customer's life easy. Therefore, the idea is to make it easy to place an order, to change one, to obtain service, to resolve problems and issues that arise. Easy! That's going to be the name of the game in the future. With all else being equal, the company that makes it the easiest for the customer to do business is going to be the leader in its respective industry. It's that simple. What can be difficult is getting there.

Management Commitment: Many company-sponsored initiatives, under the guise of being a primary company objective, lack a collective management commitment to the process. Generally, a select few are charged with pushing an initiative, in many cases before all the major players understand it fully or participate in its development. Many, inside and outside management, perceive these initiatives as the "program of the year." The expectation is that it will eventually be replaced with yet another program of the year. Before long, employees simply do not take the company seriously.

In pursuit of such initiatives, management often tells employees they must become thoroughly committed to the process. Often it isn't long before a perception is established that management obviously meant that to apply to everyone but itself. I've yet to see a case where an initiative was started without good intentions, but if this kind of perception is created, the average middle manager and production associate will not be convinced of its importance. Therefore, they will only do what they are required, in order to show cursory support.

WFE requires an extensive commitment on the part of the entire management team; that must be well documented, well communicated, and continually reinforced. WFE should not be entered into lightly because it should be something that is difficult to stop, regardless of how or how often the players themselves may change.

In essence, this particular commitment is meant to represent the company and its desire to serve the customer better than any competitor, now and in the future. If the company's senior management is not ready for such a commitment, then it should forget about lean engineering. Why? I believe I can best address that question by asking a few more:

Why make change in manufacturing that could reduce lead times, improve quality, increase flexibility, etc., if management has no intention of fitting this into the overall business scheme? Why not pursue the same type of improvements achieved in manufacturing in other facets of the business? And, last but not least, why even start a lean engineering initiative if the company has no intention of taking it to its ultimate level of accomplishment?

I could go on, but I've made my point. Assuming, however, that management sees the need and the opportunities provided through the pursuit of a waste-free enterprise, a company must focus on three things.

First, a company or corporate-sponsored function has to be organized and set forth. The company must show it plans to pursue this process throughout the entire enterprise. Management, therefore has the responsibility to organize, prioritize, outline, and communicate what the scope, purpose, and intent of this function will be as well as the responsibilities they will ask employees to carry out. This means members of this function will have to be highly qualified communicators and instructors and will have to be well versed and highly committed to the lean engineering process.

Second, the company must create an effective means for auditing, measuring, and reporting progress for senior management and the entire workforce. This calls for new measurement principles and a number of added, but different, measurements specifically designed to track progress.

Finally, serious incentives must be established. Most often, this can call for a profit-sharing venture. The key is to find what the workforce values. This could range from time off the job to tuition reimbursement for higher education. The employees must view these as important, otherwise the company is asking the workforce again to take up arms and proceed into battle for a sizable war on waste and inefficiency with little if any incentive to do so. In this case, the "warriors' efforts" would likely be less than stellar.

Employee Participation: I'm referring here to an extended level of employee participation. If the waste-free enterprise is to become a reality, the employees must commit to the process across the entire enterprise. In fact, total participation is necessary because every employee has an important role to play in this venture.

When all is said and done, people are the key factor in what makes or breaks any initiative. So, the key to success rests with gaining the interest and support of all employees. Doing this starts by helping them understand nothing can be taken for granted. You can be first today and last tomorrow in the race for success and, metaphorically speaking, it's easy to describe.

We can use Futurist Joel Barker's thoughts on paradigms as a guide. Barker tells us when a business is capable of serving the customer better than anyone else in the industry, the timepiece of success automatically reverts back to zero. It's then a new race for everyone. However, remember the competitor who caused this shift is already out of the blocks, off and running. Obviously, starting again, along with giving a sizable handicap to a strong competitor is a difficult race to win.

The answer to competitive leadership, of course, is to get in front of all the other runners and stay there. And, though the WFE (or any other lean engineering effort that encompasses the entire enterprise) may not be the panacea, it can be a potential problem or a golden opportunity. The potential problem is that if you aren't doing it, your competition can and may be. The golden opportunity is that you could be the first off the blocks with such an initiative. Either way, it can only be accomplished with the active support of the workforce. Conversely, it can be easily damaged, if not destroyed, by uninformed, underutilized, and disinterested employees.

Supplier Involvement: A company pursuing a WFE doesn't have to demand its suppliers are Lean although this would be preferable. However, it must have suppliers who are involved with the process and committed to supporting it. This calls for them to have an appreciation for what the WFE-oriented company wants to accomplish. Primarily, as this pertains to suppliers, the objectives are outstanding quality and flexible delivery. Once a company can say it has achieved this with its suppliers, it has set the stage for serving their customers better than all others. If the WFE company has all the other elements of the process up and working but has suppliers who cannot meet these two essential objectives, the total effort will fail or be seriously weakened.

Achieving the appropriate level of supplier involvement and commitment to the process doesn't, in my judgment, mean demanding they follow your lead with WFE. Some of them will never take their improvement process that far. Though you can say that if they don't you'll take your business elsewhere, odds are you will not always be able to find an adequate replacement. As a result, if a company makes such a demand and retreats the first time a supplier doesn't commit, you will have a difficult time explaining to your entire supplier base why it was required of one but not others.

Therefore, the objective becomes to establish a set of guidelines and expectations, which all suppliers can achieve. For example, one of the established guidelines could be that you will give all suppliers a one-week "firm," plus a three-week "flexible" order schedule. This means they can depend on you accepting the products produced under the firm schedule. However, they must be willing to accept changes in quantity and mix for, say, the third week of the flexible schedule and some major changes in week four. Accordingly suppliers who are into lean practices could meet these guidelines with little or no problem. Others, not into the lean approach, may have to run and store inventory to fulfill this obligation and take the chance of scrap, rework, or obsolescence. However, both could meet the requirement without an immediate and expensive overhaul of their processes and systems.

Of course, one of the things a good WFE company will have in place is something designed to help its suppliers who are not using Lean practices. The obligation on the part of the WFE-oriented company should be to aid its suppliers in becoming educated in lean practices. In most cases, they are willing to accept the help and advice offered.

Process Discipline and Continuous Improvement: Today, it is common for management to expect manufacturing to make Lean improvements continually. For the successful enterprise, however, the mentality starts to shift toward being lean in all the other business areas. Sales, for example, has to focus on driving waste and inefficiency out of its operation and has to feel an obligation to continue to do this forever. But so does Marketing, Engineering, and every other major function in today's modern business. Even support functions, such as Human Resources, Finance, and Quality Assurance should not be immune to this duty.

The basic tools and techniques of WFM can be applied to the office environment as well as the production floor. However, for those who thought they had their hands full with resistance to change on the shop floor, believe me, it is nothing in comparison to making change in the office arena. Based on my own experience, for every degree of resistance you experience on the shop floor in implementing significant change, you can multiply that by a factor of ten when you're making that type of change in the office.

This is an unfortunate yet simple truth. The same professionals who are typically critical of manufacturing—who push so hard and energetically on the need for manufacturing to accept change and "just do it"— are the same people who often fight tooth and nail to keep their areas status quo. Allow me to relate a little story to emphasize the point.

I came up in industry through the Industrial Engineering ranks. For anyone who may not know Industrial Engineering, it primarily focuses on more efficient manufacturing methods. In the process, Industrial Engineers measure the time and motion of operators. Back in my day, they used stopwatches to accomplish this. As a young IE always looking for a better way, I found a new digital stopwatch, which was more accurate than the old hand-wound sweepers. When I took this to my boss and suggested we purchase a few, do you suppose he greeted the idea with enthusiasm and acceptance? Far from it. In fact, I was chastised for bringing it up and was told the watches we were using were fine. Today, the old sweepers are gone, replaced with digital technology. This is because they were found to provide much more precise readings, which is of major importance in performing time and motion analysis.

But consider this. Here was a function designed to make change (and who made a practice of preaching the importance of it to others) who was as quick as any other function to fight change when it came down to their area.

I can assure you things haven't changed much. Professionals who work in the "front office," as many production associates call it, have a special way of thinking when it comes to change. Unfortunately, this frequently boils down to the feeling that while change is needed on the shop floor, in the employees' particular area things are "perfectly fine the way they are."

I am not being critical of any office function. The facts are (just as with manufacturing personnel) each section of the business has to be given the opportunity to see that vital opportunities for continuous improvement exist in their areas. It is then a matter of understanding the tools required to achieve this end. Nothing more, nothing less.

In doing this, what they will find is they can "flow chart" activities in their functions to identify wastes almost as easily as flow charting a standard production process. They will find opportunities, within certain limits, for Workplace Organization, Uninterrupted Flow, Insignificant Change-Over, and Error-Free Processing. However, I would be remiss if I did not point out the limitations are usually small.

Now, a word about the importance of maintaining the disciplines of the process. One of the major problems I have witnessed in countless industries is a total lack of discipline. I often see production processes improve dramatically only to slip back into old practices. What a terrible waste of good time, effort, and money. This is especially true when the answer to maintaining good discipline is as simple as establishing ongoing management audit and reviews of the process.

Achieving the waste-free enterprise reduces to helping employees clearly understand that making change and maintaining it is equally important. If there is a place in the Lean process where management should be disappointed (and, yes, possibly even express some controlled anger), it is when a team has allowed a process to slip back. This slip could be something as major as discovering machines have been arranged in opposition to "One-Piece Flow;" or as minor as finding the wrong screws in a container labeled for another. Management should view both these examples with equal seriousness. If there are going to be management repercussions, these should be applied with equal force in both situations. Both are serious because the scope isn't anywhere near as important as the extent to which disciplines are failing.

THE NEED FOR FOCUS IN OUR SERVICE SECTOR

I would like to point out the need for the same type of focus in our national service sector, the fastest growing segment of the entire business chain in America. As reinforcement of this need, all anyone has to do is to look back at the airline industry and other such high-use, high-exposure, service businesses. Customers aren't happy. Take a trip, rent a car, stay overnight in a motel, go out to eat, visit your nearest department store and more than half the time, you will be a dissatisfied customer.

Why is this the case? It boils down to industries who do not understand or care about what their customers value. It happened in the manufacturing sector before the wake-up call in the mid-'70s and early '80s. From cars to everything else, the focus at the time was to sell, sell, sell, regardless of quality and reliability. Manufacturing's attitude was if the products offered were not preferred or failed to meet expectations (e.g., they broke sooner than expected or wore out entirely), consumers had two choices: Take what you have and be glad to have it or go out and buy a new one. Of course, all this started to change when the Japanese came on the scene with automobiles and other customer-oriented products. Product variety, quality, and reliability became the name of the game and those industries unwilling to adapt to this change fell by the wayside.

Since I don't have a crystal ball, I can't predict when and how the same will take place in the service sector. But I am certain it will happen. A company, or industry for that matter, can ignore its customers only so long. In the end, someone will come on the scene who understands customers' values and find a means to satisfy them.

All the talk, energy, and effort expended nationally over the past ten years at the manufacturing level, (to become globally competitive and serve the customer better) will pale in comparison to the re-engineering that will occur in the service branch. The change is inevitable because the same positive benefits achieved through the lean initiative, in the manufacturing industry, will be demanded in the service sector. All that customers are currently waiting on is those competitors who are wise enough to put such practices in place. When they do, those competitors will capture their respective markets.

THE WARRING ADVENTURE, PART 7

Jeff Simpson's initial meeting with Brooks and Owens went exceptionally well. As he later reported to Jim, after hearing their plans and the driving strategy they laid out for the company—in which DQS2000+ played a key role—he noted he concluded his session with the following key points:

1. Though DQS2000 was an admirable initiative, it did not go far enough. It focused primarily on "product" quality and failed to address quality on a broader scale including, along with product, all functions, all systems, and all people.
2. The organization that recognizes and focuses on being the easiest for the customer to deal with and initiates practices, systems, and procedures that accomplish this, will command the market share and become the primary force in the industry.
3. Finally, for any manufacturing-oriented company like Denning to have not yet incorporated lean manufacturing in all its production processes, leaves it vulnerable and a gigantic step behind at least some of the competition.

Jeff stated Owens and Brooks (particularly Owens) were concerned about any effort that would require them to step back from the DQS2000+ initiative. He said he told them that wouldn't be necessary and that the answer could be as simple as going to the workforce with an extension of DQS2000+. He even gave them a few suggestions on the matter. One would be to establish the "+" in the acronym as achieving a lean status throughout the enterprise. Another was that they might consider changing one letter in the acronym. From an S to an E, with the ending letter now representing "Enterprise" rather than "System." This would change the initiative to the "Denning Quality Enterprise" and would therefore indicate DQE was to be driven throughout the entire company.

Jeff went on to say they seemed to like the suggestions, particularly the last one, and appeared to be willing to make some adjustments in the DQS2000+ outline. He finished by saying they appeared to buy in to his ideas and that he had been requested to draw up a consulting proposal within the next two weeks. In the meantime, they told Jeff they would consider what he had said and would decide after they had studied it and the upcoming proposal.

Jim couldn't have been more pleased and told Jeff so, but shortly after Jeff departed, Jim found he was in for an even bigger surprise. Owens called and asked that he stop by her office. When he arrived, she invited him to take a seat.

"Jim, Brooks and I listened to what Mr. Simpson had to say and I want to let you know that we were impressed," she started.

"I'm glad to hear that," Jim responded.

"He's most impressive and had some good insight into what it takes to be a world-class business, as he defined it," she continued.

Jim nodded affirmatively.

"We found his general observations about our long-range strategy to be interesting, to say the least. He convinced us that we're not taking DQS2000+ anywhere near as far as we should, considering it's currently the leading initiative for the corporation.

Jim was enjoying what he was hearing but was beginning to wonder when she was going to get to the point. There was something beyond all this that she wanted to share with him. In the meantime, he kept politely nodding in response to what she had to say.

"The reason I asked you to drop by, however, was to discuss a potential opportunity for the both of us," she noted.

At last, he thought, *I'm finally going to hear it.* Owens paused only briefly before continuing.

"Jim, you probably understand better than anyone the concept Jeff is pushing. In fact, he told us you two used to work together; and he said he had nothing but absolute confidence in your basic knowledge of the lean process, as I believed he called it."

"Yes, that's correct. We called it lean manufacturing at the time," responded Jim.

"Of course, I believe you know he's pushing this to a higher level. To include the total business, I mean."

Considering she intended that as a question, he responded, "Yes. Certainly."

"Then I believe you can understand that if we decide to pursue the matter, I am going to need someone to head up the effort for me."

Jim was suddenly hoping she wasn't thinking about asking him. But he no sooner got the thought out when that was exactly what happened.

"Brooks and I discussed this matter in some detail and we both agree that you're the best person for the job," she said, stopping and smiling broadly, waiting for his response.

Jim felt like standing and dashing out of the room. The last thing he wanted to do was become a "desk jockey" in the corporate offices. He was an

Industrial Engineer by education and had spent his entire career in a hands-on role. He liked manufacturing. It was where the action was, and he especially liked his current role. He was responsible for managing a huge, important function of the business, with literally hundreds of people reporting to him. To fall back now to an individual contributor wasn't something he had any interest in doing. However, he composed himself and managed a reply.

"Are you saying you would want me to step down as plant manager?"

"Not that we would want you to. Nothing like that. It's just that we have a bigger and more important role we'd like you to play. I can assure you, if you do the kind of job I think you're capable of, this would open a lot of doors for you in the future."

"Well, I don't know what to say," Jim replied.

"Believe me, Jim, I'm not asking you to make a decision of any kind at this moment. We haven't even decided if we're going to proceed down this path. But, if we do, we want to know you could be depended on. Of course, we want you to understand that we would structure this position in a manner you could feel comfortable with. In essence, what I'm looking for right now is a commitment that you would at least consider it. We don't have anyone in the executive ranks who knows anything about the lean process other than you, and we're currently on a fast track. Frankly, we just don't have the time to recruit someone to fill the role. So, it's a simple matter of filling the job internally or forgetting about it for the time being."

"I see," replied Jim, finding it difficult to make an argument or proceed with further discussion on the matter.

"Well?" said Owens.

Jim thought before taking a deep breath and allowing it to escape from his lungs slowly. He straightened himself and then leaned back. "All right. You can depend on me."

Owens smiled again, stood, and extended her hand to him. "Great," she said as they exchanged a firm handshake.

A WILLINGNESS TO ACCEPT THE CONSEQUENCES OF YOUR CONVICTIONS

One of the last things I want to address, regarding leading effective change and especially leading the lean engineering initiative, is making certain you understand the potential consequences of your convictions. This isn't a bad thing, but it's something you need to recognize.

Remember the old adage: Be careful what you ask for because you might get it! In some small measure, this seems to be true for those who become strong ambassadors of the lean process. When they get the appropriate people's attention by so energetically pushing the process, they are often asked to play a major role in implementing it. Typically, this is because if they are successful ambassadors, then they are reasonably good communicators and salespeople. Therefore, they are just the kind of people needed to help lead and direct the workforce. In addition, the strongest ambassadors of the process usually have some prior experience with implementing it elsewhere.

I have seen this happen repeatedly throughout my career: If you are not interested in possibly having to change your position, if not your entire career focus, then be careful in becoming the initial punch behind a lean engineering effort. On the other hand, if you have already assumed this role, you most likely have experienced the benefits of the lean process, hold firm convictions about its need, and would therefore be willing to play an active role in its implementation. But, this isn't always the case.

We have seen people who like to take a new idea or concept, learn enough about it to discuss its merits intelligently, and then leverage this in hopes of some special recognition and/or personal reward. They are what I call "seed droppers." This is entirely different from "farmers" who are willing to work at making something grow from seed to harvested fruit. The droppers pitch seeds around hoping someone will come along to plant and harvest them. Droppers have no interest in actively farming an idea. They would prefer to proceed to the next new theory to be used to their benefit. They tend to make a career of preaching change but seldom apply themselves to make it happen. If it sounds as if I dislike seed droppers (you will find them in almost every business), then you are right. Putting that aside, however, if you hold strong convictions about the lean process and decide to do what it takes to get everyone's attention, you should be prepared to be asked to do more because, in all likelihood, that is what will happen!

SETTING THE STAGE FOR WFE: THE SIX ESSENTIAL STEPS

Focusing again on the WFE, applying the lean concept to the entire business is a larger and more complicated task than a specific focus on the manufacturing arena. On the other hand, one of the things I believe many companies have forgotten (or haven't taken into consideration) is the need to keep the process simple. Most of the efforts I have seen have been directed at a complete overhaul of existing systems, practices, and procedures to accomplish this. Additionally, you will seldom, if ever, find a step-by-step approach to getting there, such as a list of specific tasks and in what order they should be pursued.

Being an IE by profession, I've always believed in using an analytical approach, where possible. I never ceased to be amazed how companies think they can take the ninth step, for example, before they take the first, second, and third in appropriate succession. Put another way, how effective do you think a person would be trying to win a 100-yard dash by making an effort to leap to the 50-yard line at the sound of the gun? He or she would fall on their face and look foolish while everyone else in the race went flying by.

Therefore, I am going to lay out six essential steps to get the process of WFE started in an effective manner. I believe these should be taken in the order prescribed, without jumping ahead. However, you may be able to do some of these in conjunction with others. Once you take these steps, advancing the process becomes a matter of continuous improvement, in these and other areas of the business (see Figure 8-2).

As you will see, each of these following steps cover a broad spectrum and I don't intend to provide a detailed how-to outline. Rather, the intent is to give the reader a basic understanding and the order of execution.

Step One: Commit to the Process. As with any good initiative, it all has to start with a solid management commitment. In this particular case, it needs to start at the top. In fact, be actively driven from that level. Among other factors, three important issues have to be addressed in effectively taking this initial step. They are organization, incentives, and measurement. The following briefly addresses each of these:

Organization: You must make plans for an organizational structure that fully supports the effort. There are countless ways to do this, but I can outline a simple and straightforward one. First, put together a group of up to five highly talented people well versed in lean engineering and reporting directly to the person at the top (or at least one of the key executives who report to this individual). Their job would be to roll out the process through the company. In keeping, they would establish effective measurement tools, and continually audit progress—including

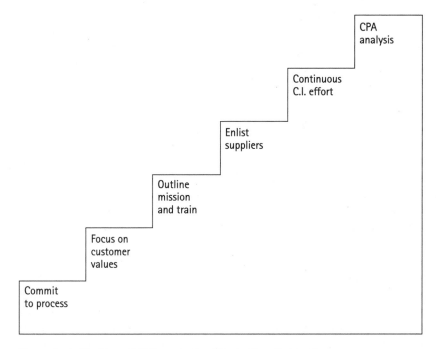

Figure 8-2. Six Essential Steps to the Waste-Free Enterprise

an audit of the customer base. Second, insist that every business division within the company appoint one or two highly qualified people to interface with this group. Their principal role would be to head up training and implementation at the operational level. In this case, these people should report to the corporate group on a dotted-line basis. Last, insist that every function (and/or department) within the divisions prepare to assign one person as the process coordinator. The individual does not have to take this role full-time but should be able to as an extended part of their normal job.

The key to the matter of organization, I believe, is to insure that all the bases are covered regarding the structural ability to forward the process throughout the entire company or corporation. It is also important to do this without creating a monstrous support organization, which can get in the way of taking care of ongoing business duties and responsibilities. While I am one of the first people who would push the need for the company-wide lean initiative, I also know the dangers in over-staffing such a venture.

Incentives: There is no set approach to what the associated incentives should be, other than it should be something employees value. What I want to address here deals more with where you should place such rewards. Outside of incentives for overall progress and accomplishments, one of the other key areas would be employee training. Special incentives should be set up where certain levels of accomplished knowledge and expertise are awarded to those who extend themselves beyond any of the standard, required employee training. This is a good first assignment for the corporate group. Be careful, however, because efforts should be made to insure incentives are not placed on what should be viewed as nothing more than standard job expectations. One of the more popular means of initiating such an incentive program is through a "Gain Sharing" approach. Here, incentives are paid based on progress against an established base, which normally includes measurements related to quality, productivity, and most important, meeting customer commitments. However, I feel most fall short in the "challenge category." The good incentive program should include measurements against goals that, while achievable, are challenging. For example, while meeting standard established lead times and/or quality levels to earn some remuneration level, steps should be built into the program allowing for added incentives when these levels are exceeded. Just as important, penalties should exist for when minimum requirements are not achieved.

Employees must understand that such an incentive program is not designed to supplement their normal income, but as a true bonus for work and effort, above and beyond some standard call to duty. This, therefore, calls for some thoughtful way of communicating the initiative's purpose and intent to the entire workforce.

Measurements: This is a crucial issue as it pertains to the lean initiative success. The company must be willing to incorporate new measurements that drive its employees to do the right things. This means some of the old standby measurements, particularly those used in manufacturing, have to be evaluated and, in come cases, put to rest. The list is numerous: "direct to indirect labor ratios," "Inventory Turns" (based on dollars rather than pieces), "equipment utilization" (based on the percentage of up-time only). These measurements lead employees toward what has been described as dysfunctional actions. (See *Fast Track to Waste-Free Manufacturing*, "Measuring Progress—What to Measure and How" in chapter eight.)

Step Two: Focus on customer values. Start here by working on the three fundamentals for good customer satisfaction regardless of the type of product or service provided: Speed, Flexibility, and Quality.

For speed, the initial effort should be on taking the manufacturing lead time advantages, gained through the lean manufacturing effort, all the way to the customer level. (You must apply the concept in manufacturing before you can even consider taking the approach forward to the entire enterprise.) Apply these improvements to the total company lead time chain and make it known to the customer. Also, further make it known that you intend to work on driving the lead time down even further and, just as important, that you intend to accelerate every business element, from order entry to delivery and beyond.

For flexibility, take manufacturing to a higher level of achievement, which is to gain the complete ability for mixed modeling scheduling throughout. This provides the initial flexibility to serve the customer better. In the meantime, go out of your way (as needed) and at additional cost (if required) to satisfy ever-changing customer requirements. Remember, the name of the game is to make it easy for your customers to do business with you and you can't well do that if you are not willing to show them you are serious about fulfilling their needs.

In the area of quality, the focus shifts beyond product quality itself into all the related activities serving to link the entire customer chain. In other words, work to improve the quality of the order entry system, the customer service branch of the business, the delivery sector, and so forth. In addition, improve the quality of overall concern and responsiveness. This means keeping an open ear for the customer, where they can be heard 24 hours a day, seven days a week. In this case, it starts by shying away from modern technology just a bit and providing a "real human" that replies when customers call. Nothing, and I mean nothing, aggravates and irritates customers as much as not being able to speak with a responsible individual. Having to wait to do this while a recording asks them to wade through a list of options, getting a busy signal or another recording, is about the most ridiculous concept anyone ever dreamed up. The one who first sold this idea to the general business world as some sort of advancement in customer satisfaction should be tarred and feathered and rode out of town on his quack wagon. We would have all been better off for it.

Step Three: Outline the mission and train employees. This step is directed at communications and training for the entire workforce. Some may wonder why this isn't listed first after management commitment. The answer is that starting with a focus on customer values and making some of the initial improvements outlined can be accomplished at management's discretion. In other words, it can do this without any formalized process to support it. In addition, doing this reminds employees management is serious enough that it has already incorporated some changes, aimed at things that are important things to any customer regardless of product or service provided.

The communications and subsequent training have three essential aspects:

1. A new mission statement aimed at outlining the company's future focus must be prepared and communicated to the workforce. The toughest task is to make the statement as short, precise, and clear as possible. Therefore, say a lot in a few words, which requires thoughtful consideration. Remember the following two hints: Since you can't be all things to all people, decide what your business forte is going to be and then decide how to get there. Though you have to focus on improvements in all business areas, you can't (in all likelihood) be the best at everything. Therefore, you have to decide where your prime opportunities for success rest and then pursue them aggressively.

   For example, a company may have good reason to believe the prime opportunity rests in its ability to provide new products and technological advancements. The questions to ask are these: Can you do this better than anyone else in the industry? If so, how? If not, why? Assuming, however, the company is convinced it should proceed with this particular focus, the mission statement could be: "To be the best in the industry in speed to market with new product development and technological advancements." Short, precise, and to the point.

   Avoid confusing the mission statement with other flowery words and poise, such as with the best quality, the best delivery, and/or with involved and inspired employees. What everyone will come to understand through training, instructions, and practice is that those other things are a given. However, in this case, the one thing employees must never lose focus of is being the best in the industry with new product development and technological advancements.

In all probability, formulating a mission statement that will be the heart and soul of the company will be more difficult than the example provided, but you get my point. Keep it direct and as straightforward as possible. Here is my best advice on this: If it ends up on a plaque posted in the main lobby, it should be something that anyone passing by could read and understand without having to stop and study it.

2. The primary emphasis in the extensive training of all employees (and, again, I mean "all") must be centered on the tools they can use in identifying the opportunities and making continuous improvement.

Most often, the principle WFE tools involved in accomplishing this are "Waste Assessment" and "Activity Flow Charting." I have witnessed business process sessions aimed at examining the office areas where participants were tied up for extensive periods of time performing "Standard Work" for every activity outlined. In my judgment, this is a horrid waste of good time, energy, and resources and isn't required. You don't need to time study every activity to allow commonsense to reach the same conclusions.

It's important to understand that the training we're talking about is not a one-time exercise. Nor should you examine each business process for more improvement. The initial training and improvement efforts should be directed at picking most of the low hanging fruit. Later sessions could call for performing "Standard Work" in order to help streamline the process further. However, this isn't needed in the beginning.

Step Four: Enlist suppliers. At this point, it is time to enlist your supplier base in the strategy, outline expectations, and assist them where possible in training and engaging their workforce in the effort.

Part and parcel to this is to establish a bona fide Supplier Certification Program—which most companies already have—and insure that the goals, objectives, concepts, and techniques of WFE are implanted in the process. What American industry will come to recognize is that it is no longer good enough for a supplier to ship quality parts on time. Suppliers must learn to do this more effectively, thus bringing down their costs and their price because the company interested in its suppliers is also interested in its suppliers' welfare. If it is interested in their welfare, it will be interested in pricing improvements achieved at the suppliers' operation, through productivity and efficiency gains and not by cutting into suppliers' profits through a demand for continuing price reductions. What I am establishing here is that a globally competitive company must assume some standing obligation to help its suppliers rather than following the age-old practice of continually "beating them into submission."

In the end, a Partnering relationship should be actively pursued and suppliers should be invited to help the company with product enhancements and new product designs. If achieved, an operation will find almost immeasurable value in this relationship.

Step Five: Make continuous improvement a continuous effort. The fifth step is one of the more important. It is to put policy and practice in place designed to assure continuous improvement is a never-ending effort by all concerned.

Again, I believe this has to start with an incentive program designed to focus on the more important aspects of the process. As mentioned, one of the more popular approaches in American industry is a Gain Sharing program. However, in my judgment, most Gain Sharing programs already in place would have to refocus their key measurements. The need is to adjust the focus more outwardly than inwardly. Typically, most companies use Gain Sharing measure with an inward focus (e.g., plant scrap rework, downtime, internal quality levels), in the hopes this will satisfy the customer more and improve profitability. But the equation's outward portion—the customer—is what counts. If you are doing the right things to satisfy your customers, the profit picture will take care of itself (assuming you aren't giving your products away in order to stay in business).

Even though I would not want to oversimplify making continuous improvement a continuous effort, this should become a natural response to the right incentives and measurements because employees perform to the level by which they are measured and rewarded.

After completing all these steps, your organization will not be a waste-free enterprise, but the stage is set to proceed more expediently to success. This happens because management has made a commitment, backed by an appropriate organizational structure and incentives highly valued by employees. In addition, effective measurements have been established and suppliers have been actively engaged in the process. Finally, continuous improvement has become policy rather than some desire of the company. You might describe accomplishing all this as a fast dash in the right direction without falling on your face in the process.

Step Six: Conduct a serious Core Process Analysis (CPA). After hearing what this last step is all about, some might argue it should be first. However, it is more important to begin the process with what you have in place and then take steps to improve it. If a company decided to conduct a serious CPA before it started taking the other five steps, WFE would probably never get off the ground.

I spoke briefly about CPA in my first work, but now it is important to explain how it works. To begin, two guiding factors drive the entire CPA process: Uninterrupted Flow and Manufacturing Expertise.

Uninterrupted Flow: Everything about WFM is structured around the principle of Uninterrupted Flow. As many manufacturing operations are putting the techniques and concepts of WFM and other lean engineering processes in place to improve flow, they are not achieving the ultimate. This is because they have monuments within their operation, which will not allow it. What I am highlighting here regarding monuments is not what is typically classified as such. Of course, the ultimate in Uninterrupted Flow would be to have the equipment required to perform value-added work flexible enough to move, at will, from job to job.

Even operations that have done massive work on One-Piece Flow and other Kaizen-related activities, will most often require some rather extensive controlled batching, or what is usually classified as Kanban applications. Typically this is for large processes such as painting and/or plating facilities. In reality, any machine in your factory that does not have the ability to move, at will, from point to point, is in the truest sense a monument. While achieving the ultimate ability noted is beyond most manufacturers' ability, it is something they should consciously recognize as a road-block to continuous flow.

An analogy I used when training others was to ask them to consider a group of carpenters who, in order to build a house, were required to take their work to a number of fixed tool stations rather than having tools they could freely move with them (hammer, handsaw, etc.). Imagine a "Nailing Station" where anything needing nails had to be brought for work to be completed. Then a Sawing Station and a Painting Station and so on. In some cases the carpenters would have to move the entire house in order to perform the work required. Now, while this example serves to over-exaggerate, this is precisely the way the typical manufacturing engineer designs most production processes and the way batch manufacturing is essentially conducted.

Usually the emphasis on manufacturing equipment design is on bigger and more complex rather than on smaller and more flexible. Regardless, to be a globally competitive manufacturer, flow has to take the highest priority. Therefore, under the CPA format, analyze every factory process using the following basic criteria:

- Does the process obstruct Uninterrupted Flow. and if so, is it critical that you perform the work in-house?
- Can you make the process smaller and more flexible without an unacceptable expenditure?
- Does the process perform value-added work in the customer's eyes? (Value-added work is something the customer is unquestionably willing to pay for and/or specifies must be done.)

Manufacturing Expertise: This has to do with whether the work being performed is an expertise required to be competitive. CPA asks the fundamental question as to what a company should be doing to reach the highest achievable level of competitiveness. As important, however, is the analysis of what a company should not be doing. This comes down to examining what I call the "heart" of the product produced. As an example, the key components of an air conditioning system are generally viewed by the industry to be the compressor and the coil, thus what I refer to as the "heart" of the product. Therefore, companies, which purchase their compressors and/or coils rather than design and build them in-house, are at the competition's mercy. This is because the competition generally sees holding the ability to design and manufacture these components as essential for market leadership and longstanding competitiveness.

My experience has been that most manufacturers who have been in the business started out with about the right processing but then tended to expand their overall expertise as time went by. Vertical Integration was a popular approach in the 1960s and 1970s, and most everyone followed that prescription. Many firms set up processing in their factories for work they were ill-prepared to perform. It was often a springboard for poor quality, field failures, poor customer satisfaction, along with over-staffed, slow, and cumbersome manufacturing.

A company should ask itself if it has taken on too broad a spectrum of required expertise to produce and deliver its products. Unfortunately, a company cannot be a manufacturing expert at everything and should remember that for every new complex manufacturing process it assumes, it effectively establishes another potential bureaucratic organizational level. If it is serious about adding new forms of manufacturing, it will hire talent and/or establish functions with the qualified expertise necessary to do the job. What happens, however, is that the company stretches the expertise of its existing workforce beyond its ability to maintain a good competitive focus on any one thing. As this occurs, it pays less attention to staying abreast of the latest technology, and any clear-cut competitive advantages that once existed begin to deteriorate.

This is like assembling a professional hockey team and then insisting it be a champion in hockey *and* in baseball, football, and track. We know this would be unlikely, if not impossible. In fact, assuming the team had championship-caliber players in hockey, if they had to focus on and be competitive in other sports, their expertise in hockey would begin to suffer. Unfortunately, we are often guilty of doing just that in our manufacturing operations because we think nothing of asking employees to become specialists in yet another field of manufacturing endeavor.

The basic criterion for examining this factor is to ask this: Could work that is not considered to be a critical competitive expertise be done outside, at a cost equal to or perhaps slightly higher than current standards?

That's correct, at a *slightly higher* cost. Remember that standard costs on many of the more complex processes in no way reflect the actual cost of being in the business. While this, of course, has to be analyzed, it is important not to allow standard accounting practices to lead you down the wrong path. However, in almost every situation where I have pursued outsourcing, we were able to come back with pricing lower than our current cost standards.

In making a genuine effort to keep Flow and Expertise in mind, a CPA evaluation will lead a company to identify specific improvement opportunities. Acting on these opportunities, however, becomes the key to success. If done properly, a company can turn a slow and cumbersome manufacturing operation into a lean, fast, and more effective competitive weapon.

Jim mentally wrestled over the next couple of weeks with Owens' words. One side of him felt excited and encouraged about the opportunity. The other kept coming back to his giving up a plant manager job for, more or less, an individual contributor position.

He went so far as to call Jeff Simpson and inquire if the subject had, in any way, surfaced during his conversation with Brooks and Owens. Jeff assured him it hadn't but wasn't in the least surprised he was being considered. He then went on to assure Jim he thought it was an excellent idea and that he would be pleased if it happened.

Nothing said by Jeff or any discussions he'd had on the pros and cons with Ginger, helped him reach a conclusion. He was convinced it was going to be one of the tougher career decisions he'd ever made.

On the business side, things seemed to be going from bad to worse. He was facing a severe cutback in resources if something didn't break for the better soon. What made it worse was that he had only recently brought back a good number of production associates in anticipation of the upcoming "busy season" for the business. As it now stood, things didn't appear to be all that busy. While Sales and Marketing were working frantically with Aklin and SJM Industries to salvage the business, the feedback was not that encouraging. Jim had the manufacturing side working overtime to build the product mix needed by Aklin, which was expensive and was fraying employees' nerves.

He felt that timing couldn't be worse to be considering a significant shift in focus for DQS2000+, coupled with changing the leadership in manufacturing, and starting a whole new, expanded venture into lean engineering. He absolutely cringed every time he thought about walking away from manufacturing without having achieved what he set out to do and, more important, turning it over to someone else with all the problems at hand. He would feel like an absolute failure.

He needed good advice, so he again turned to his friend and mentor, Frank Zimmer. They ended up having dinner at the country club where Jim and Ginger held a membership. After the meal and some relaxing conversation, Frank finally opened the door by asking Jim how things were going.

Jim took the time to cover Jeff's recent visit and the situation he had subsequently been placed in. He openly admitted: "I'm having a hard time with this."

"Well, I can see why," responded Frank. "That would be a big decision to make. Give up the job of plant manager to take on some sort of corporate advisory position? That's a tough one."

Jim was a little disappointed Frank's response wasn't one of his usual advisories, aimed at all the good reasons he should be jumping to accept the job. But that wasn't to be.

"Who have they got in mind to take your place?" inquired Frank.

"Beats me. It isn't something the boss and I have discussed."

"Really? I would think that should be at the top of the list," Frank noted, pausing.

"You know, Jim, it sounds to me like your boss doesn't think it would be any problem whatsoever finding someone to take your place. If that's the case, that's an issue in itself. True enough, you may be the only guy to take the lean initiative and run with it, but you're also the guy in charge of manufacturing. Still, the both of you should know there would be someone taking your place who would keep things on the right track."

"I'm sure her answer to that would be to recruit someone with experience from the outside," Jim replied.

"Fair enough. But that's going to take some quality time and it sounds like she wants you in the new role almost immediately, if and when a decision is made to head that way. All I'm saying, my friend, is make sure of what's on her mind."

"I'll do that," Jim responded.

"And one other thing, unless this new role was something I were interested in, I would be racking my brain for some other likely job candidates," Frank concluded.

On the way home, he thought about their discussion, but from a slightly different viewpoint than Frank had mentioned. But one thing he said was certainly true. If Jim wasn't interested in taking the position, then he needed to be prepared to offer some options to Owens. In addition, he convinced himself he needed to put aside his concerns about the plant manager position long enough to look objectively at the pros and cons of the role she was proposing. It was only the right thing to do.

The next morning, after surviving the hustle and bustle of the daily reporting activities and production meetings, he sat quietly at his desk and thoughtfully listed the pros and cons he had decided to address on his way home the previous evening.

After over an hour, with only a couple of telephone calls as slight interruptions, he stopped to study his work. The first thing that caught his eye was that the list of pros were over twice as long as the cons. He wasn't sure if this meant anything, so he decided to apply another technique. On a scale of one to ten, he ranked each item, in terms of overall business importance. When he finished and totaled everything up, the score was 130 to 80, pro versus con.

Jim was thinking if the exercise had any merit at all, the new role was more important to the company than keeping his current job. The natural conclusion would have to be if he had any doubts about it, it was because his doubts were more of self-centeredness and pride than anything else.

He admitted he had enough pride and ego to believe it would be difficult to find a replacement. But he was also smart enough to know that finding a plant manager would be easier than recruiting one with the understanding, experience, and commitment he possessed in lean engineering. And that did not even include the business knowledge he'd already gained of Denning, its products, processes, employees, and customers.

He finished his exercise by firmly deciding that if the new role met his basic expectations in terms of scope, responsibility, and duty, he would consider accepting the job.

PUTTING IT ALL TOGETHER—PART 1

I believe it is important, at this point, to pause long enough to recap WFM, WFE, and their appropriate linkage.

Waste-free manufacturing (WFM) is a process designed to address the manufacturing shop floor and to make it more efficient and customer oriented. Pursuant to this, the major difference between WFM and other lean manufacturing techniques is in how and when to use the commonly recognized tools of the trade and when and where to place a specific focus. The intent, of course, is to provide a step-by-step approach to achieve a globally competitive manufacturing status, in a reasonably short period of time (i.e., "Rapid Insertion" of the process).

Waste-free enterprise (WFE) is a process specifically intended to address the entire business chain. However, it also focuses attention on applying many of the same tools and techniques of WFM on the administrative side of the business. Most important, the entire process is driven on better understanding what is essential to the customer—their values—and reacting accordingly.

Appropriate Linkage: Incorporating WFM will greatly improve manufacturing and thus operating costs and overall efficiency. However, it will do little to serve the customer better than before unless the benefits gained are somehow built into the total business process, from order to delivery. For example, if a manufacturing area works to cut its lead time substantially and this isn't built into the standard order entry and delivery process, the customer will see little improvement, outside of perhaps better meeting established delivery commitments. While improved delivery performance to established time-lines is helpful, it isn't good enough because it doesn't cut the standard, established lead time. At best, this will only allow the company to gain some positional parity with the competition.

Further, to initiate practices in one business segment (the manufacturing shop floor) that focuses on better serving the customer and then fails to actively seek the same accomplishment in other areas (the office and general administration) is a terrible waste. Since everything addressed thus far in both of my works has centered on the issue of waste, it would be totally inappropriate for me to ignore this issue.

The key is for a company to put the right practices in place on the shop floor (WFM) and quickly follow this with an expanded process (WFE) that covers the entire business chain of events. It is that simple as well as sometimes that complicated. The latter mentioned usually comes from pursuing some shotgun approach to resolution, which can serve to confuse and therefore complicate the matter for everyone.

Twenty-six months and fourteen days later, Jim was relaxing at home on a bright Sunday afternoon in early October. Because the day was so nice, he and Ginger had decided to cook out. As the old rusty, yet still effective, gas grill warmed under the glow of recently fired coals, he had taken the time to stretch out on a nearby lounge chair. As he lay back and folded an arm under his head as a makeshift pillow, he was thinking how good it felt to be back home for a few days.

Alicia, now three years old, played just beyond him with Browser, a Hines 57 variety mutt he and Ginger had picked up at the local kennel, who believed he was the true ruler of the Warring domain. Alicia's four-month-old brother,

Aaron, was cradled in Ginger's arms as she sat opposite Jim in a matching lawn chair.

What an adventure, he thought and then caught himself wondering why he had thought of it like that. But, he told himself, that was exactly what the last four and a half years of his career had turned into. The Warring Adventure, he summarized, silently chuckling at the thought. In essence, he had gone from a first-time plant manager, struggling to set a course for the factory he was in charge of, back to the role of individual contributor as a corporate trainer and, finally, six months ago, to Vice President of Manufacturing for Denning Corporation.

He was thinking about how it all happened and to some degree marveled that it had turned out the way it did. He loved his new job even though the excessive travel he had started when he took on the role of a trainer hadn't eased much since he accepted the VP role, as Patricia Owens' replacement. Under Jim's initial leadership and the team he had been fortunate to put together, using the principles and concepts that Jeff and his consulting firm had helped him develop, manufacturing throughout Denning had made enormous strides. Jim and 12 other select individuals had circled the globe, teaching the Denning Production System, as it came to be called. This was because, when it was all said and done, Owens couldn't bring herself to change DQS2000+ to include anything other than what she had originally intended it to be. Brooks, on the other hand, had come to see the vital importance of the process Jim was leading when it became more than evident the rewards were boundless. After less than a year into the initiative, Brooks had pulled Jim out of Owens' organization, wisely perhaps, and had Jim report directly to him. Patricia, although outwardly supportive of the process, pulled further away as time went by to keep an active focus on her pet project, which of course was to push DQS2000+ aggressively forward as the grand solution for Denning. However, it wasn't long before a serious conflict over resources developed.

Jim was stirred from his thoughts as Ginger reached over to touch his arm and remind him he still had a job to do in grilling the burgers.

"They look as if they're ready," she said nodding toward the cooker.

Jim struggled himself out of the lounge chair and reminded himself he needed to keep an eye on his weight. But not today, he thought. He was going to enjoy a nice, juicy hamburger and a beer or two for a change.

As he stood over the cooker, preparing the burgers, his thoughts drifted back to the job again. Poor Patricia. He had liked her and although she had landed on her feet when she left Denning with an executive post in another company, similar to the one she had had, he felt sorry for her. She had genuinely been interested in taking the company on to better things. There was no question about her commitment or desire. However, she had allowed herself to get so wrapped up in an initiative she felt compelled to lead and guide that she was unable to see the forest for the trees. Because of Brooks' ever increasing enthusiasm about the Denning Production System, which was rapidly producing results, Patricia's firm position on the matter of DQS2000+ and inactive participation in the process pulled her and Brooks further apart. In the end, Brooks apparently saw no solution other than cutting the cord, allowing her to move on.

Initially, Jim had been appointed as a temporary fill-in for Patricia out of necessity while a search for her replacement was conducted. After four months, he and Brooks had established an effective interface, and Brooks was impressed with the leadership Jim was bringing to the process. He finally offered Jim the job and Jim jumped at the chance of moving up to the next level. Especially in a role where he could have a major impact on leading the manufacturing sector, which was his foremost desire.

Now being in the position of having Brooks' ear, his next step was to extend the process beyond manufacturing, to take full advantage of the gains all the way up the business chain to the customer.

He drifted once again out of his thoughts about work and back to the task at hand. "Burgers are ready!" he announced as he pried them off the smoky grill and placed them neatly on a platter nearby.

PUTTING IT ALL TOGETHER: A FINAL WORD

I believe successfully leading the lean initiative reduces to three key considerations:

- Knowledge as to the primary tools and techniques and how to use them.
- Foresight as to the obstacles that will surface and how to respond to these.
- Focus as to following the guiding principles through it all.

To lead an effort successfully, there must be acquired knowledge of the tools and techniques and, as important, how and when to use it. (See *Fast Track to Waste-Free Manufacturing*. Here, I pointed out a detailed, step-by-step approach which, if followed, has had a proven track record for a rapid and effective insertion of the process. I ended the book by advising: Just do it—but with knowledge.)

The knowledge referred to is best when it comes from actual experience. I would be the first to tell anyone who has been appointed for such a task or has decided to do it on their own that they should, if nothing else, obtain some practical experience. This can be achieved through special training and/or by having someone (e.g., a qualified consultant) come aboard for a time to guide and direct a pilot effort in select areas of the factory. But, it is important for the inexperienced leader to participate in this actively and gain some hands-on knowledge of the process.

Next, there is the need to acquire some appropriate foresight about the obstacles the leader will encounter as the process continues. You have to be realistic and understand that in a manufacturing operation, the person perceived responsible for such a significant level of change is always going to be the plant manager. This is whether they want it or not. All others will look to them for guidance, direction, and support of the process. If it isn't there, the process will die on the vine.

Even for those plant managers who know and accept this, running a manufacturing operation has many distractions diverting time, energy, and effort from any initiative chosen. I have outlined these distractions and how to deal with them effectively, which brings us to the subject of focus.

Referring to my first work, I stated that "Knowledge is the substance of focus." Nothing could be truer for almost anything one chooses to accomplish though it is vitally important in pursuing a lean engineering initiative. Though others in the organization should be working diligently on the process, the plant manager must be able to maintain an appropriate focus on the initiative. Otherwise, employees will begin to perceive it as just another program of the month.

One of the best practices for the leader to use in gaining knowledge through focus is to set up a standing meeting, once a month, to review the initiative's progress and nothing else. Participants should be asked to go over assigned process measurements, accomplishments, plans for the future, problems getting in the way of progress, etc.

The second practice is to qualify your decisions so they fit well with the established principles of the process. Again, in WFM these are Work Place Organization, Uninterrupted Flow, Insignificant Change-Over, and Error-Free Processing.

Here, every major decision made should include thoughtful consideration of any aspect of the decision that goes contrary to the principles.

If it does, then rethink the decision before announcing it. If it doesn't, proceed forward.

WARRING SCORECARD, PART 7

A Final Assessment

In the end, Jim did well enough as plant manager to win the opportunity to have a strong influence on the entire manufacturing arena at Denning with his promotion to VP of manufacturing. Though his personal progress in slightly over four years would have to be viewed as highly successful, the real measure of his success for—a plant manager—centered on what his work accomplished for the company. By this particular measure, Jim was slow off the starting blocks and, thus, progress was substantially delayed. While the company was progressing through its Denning Production System, which Jim helped develop and spread, it still wasn't where it should have been in setting the proper foundation for continuous improvement and greater customer satisfaction. And this was after Jim had been on the job for four years.

Though I would admit plant managers, especially new ones, cannot usually influence the overall company/corporation strategy, they can and should actively and energetically pursue lean manufacturing practices in their own operation. I can assure anyone that there are few if any companies who, in this day and age, are going to restrict such an effort. On the other hand, there are always going to be those personal and professional distractions that take a plant manager's focus away from this effort. The important thing is to recognize this and respond accordingly.

Had Jim taken more responsibility and authority upon himself to implement lean practices aggressively, he could have avoided many of the problems that surfaced. Just as important, perhaps, convincing the senior management of Denning would have been more a matter of demonstration than salesmanship. I say this because once a company sees what the Lean effort brings to the manufacturing arena, they need little else to convince them to use it throughout the company.

Jim, like many new plant managers, spent precious time with positioning tactics, for countless problems, issues, and circumstances. Though common, it can be time-consuming and distracting. If Jim had had previous experience in the role, he would have had more confidence to pursue and remain focused on his initiative. Because of this typical scenario, I've centered this instructional work on the new plant manager, but believe what I have addressed can help any plant manager, new or experienced.

Being a good plant manager is a challenge because, if done properly, the job requires walking a theoretical tight-wire strung between the top decision-makers of the company and workers on the shop floor. From a pure job perspective, the good plant manager does not perceive his or her position as being singularly "management." This is because, in reality, the position is best designed to establish an appropriate conduit between top management and those charged with producing the products a company sells and services. As a result, the required job loyalties are almost equal in both directions. While plant managers must be loyal to the company's goals, objectives, and aspirations, they must be equally loyal to those working and performing in manufacturing. All the more reason that good plant managers are not that easy to find.

Leading any important business initiative takes hard work and dedication. Successfully leading a lean engineering effort will take all this and more. So for those about to set out on their own adventure (playing a key role in making and keeping America globally competitive), I extend to you my admiration and best wishes and ask that you keep one important thing in mind at all times. This is to avoid the tendency to let the common distractions of the job stand in the way of the knowledge and focus required to make the lean initiative a solid reality.

Index

Books from Productivity Press

Productivity Press publishes books that empower individuals and companies to achieve excellence in quality, productivity, and the creative involvement of all employees. Through steadfast efforts to support the vision and strategy of continuous improvement, Productivity Press delivers today's leading-edge tools and techniques gathered directly from industry leaders around the world. Call toll-free (800) 394-6868 for our free catalog.

The 5S System
Workplace Organization and Standardization (video)
Tel-A-Train and the Productivity Development Team

The 5S System is a seven-module video-based training program in which you'll learn how to apply the 5S System through hands-on implementation and application activities. The set includes seven tapes supported by a ull- length Facilitator's Guide and Participant's Guide that provide summaries of the tapes, structured application activities, practical worksheets and checklists.
$1,995.00 / Order 5SV7
Introductory tape only / $495.00 / Order 5SV1

Becoming Lean
Inside Stories of U.S. Manufacturers
Jeffrey Liker

Most other books on lean management focus on technical methods and offer a picture of what a lean system should look like. Some provide snapshots of before and after. This is the first book to provide technical descriptions of successful solutions and performance improvements. The first book to include powerful first-hand accounts of the complete process of change, its impact on the entire organization, and the rewards and benefits of becoming lean. At the heart of this book you will find the stories of American manufacturers who have successfully implemented lean methods. Authors offer personalized accounts of their organization's lean transformation, including struggles and successes, frustrations and surprises. Now you have a unique opportunity to go inside their implementation process to see what worked, what didn't, and why. Many of these executives and managers who led the charge to becoming lean in their organizations tell their stories here for the first time!
ISBN 1-56327-173-7/ 350 pages / $35.00 / Order LEAN

Productivity Press, Dept. BK, P.O. Box 13390, Portland, OR 97213-0390
Telephone: 1–800–394–6868   Fax: 1–800–394–6286

Corporate Diagnosis
Setting the Global Standard for Excellence

Thomas L. Jackson with Constance E. Dyer

All too often, strategic planning neglects an essential first step and final step-diagnosis of the organization's current state. What's required is a systematic review of the critical factors in organizational learning and growth, factors that require monitoring, measurement, and management to ensure that your company competes successfully. This executive workbook provides a step-by-step method for diagnosing an organization's strategic health and measuring its overall competitiveness against world class standards. With checklists, charts, and detailed explanations, *Corporate Diagnosis* is a practical instruction manual. Detailed diagnostic questions in each area are provided as guidelines for developing your own self-assessment survey.
ISBN 1-56327-086-2 / 115 pages / $65.00 / Order CDIAG

Fast Track to Waste-Free Manufacturing
Straight Talk from a Plant Manager

John W. Davis

Batch, or mass, manufacturing is still the preferred system of production for most U.S.-based industry. But to survive, let alone become globally competitive, companies will have to put aside their old habitual mass manufacturing paradigms and completely change their existing system of production. In *Fast Track to Waste-Free Manufacturing: Straight Talk from a Plant Manager*, John Davis details a new and proven system called Waste-Free Manufacturing (WFM) that rapidly deploys the lean process. He covers nearly every aspect of the lean revolution and provides essential tools and techniques you will need to implement WFM. Drawing from more than 30 years of manufacturing experience, John Davis gives you tools and techniques for eliminating anything that cannot be clearly established as value added.
ISBN: 1-56327-212-1 / 425 pages / $45.00 / Order WFM

Productivity Press, Dept. BK, P.O. Box 13390, Portland, OR 97213-0390
Telephone: 1-800-394-6868   Fax: 1-800-394-6286

Implementing a Lean Management System

Thomas L. Jackson with Karen R. Jones

Does your company think and act ahead of technological change, ahead of the customer, and ahead of the competition? Thinking strategically requires a company to face these questions with a clear future image of itself. *Implementing a Lean Management System* lays out a comprehensive management system for aligning the firm's vision of the future with market realities. Based on hoshin management, the Japanese strategic planning method used by top managers for driving TQM throughout an organization, *Lean Management* is about deploying vision, strategy, and policy to all levels of daily activity. It is an eminently practical methodology emerging out of the implementation of continuous improvement methods and employee involvement. The key tools of this book build on multiskilling, the knowledge of the worker, and an understanding of the role of the new lean manufacturer.
ISBN 1-56327-085-4 / 182 pages / $65.00 / Order ILMS

Today and Tomorrow

Henry Ford

This autobiography by the world's most famous automaker reveals the thinking that changed industry forever, and provided the inspiration for Just-In-Time. Today these ideas are re-emerging to revitalize American industry. Here's the man who doubled wages, cut the price of a car in half, and produced over 2 million units a year. You will be enlightened and intrigued by the words of this colorful and remarkable man.
ISBN 0-915299-36-4 / 300 pages / $30.00 / Order FORD

Productivity Press, Dept. BK, P.O. Box 13390, Portland, OR 97213-0390
Telephone: 1-800-394-6868   Fax: 1-800-394-6286

ABOUT THE SHOPFLOOR SERIES

Put powerful and proven improvement tools in the hands of your entire workforce!

Progressive shopfloor improvement techniques are imperative for manufacturers who want to stay competitive and to achieve world class excellence. And it's the comprehensive education of all shopfloor workers that ensures full participation and success when implementing new programs. The Shopfloor Series books make practical information accessible to everyone by presenting major concepts and tools in simple, clear language and at a reading level that has been adjusted for operators by skilled instructional designers. One main idea is presented every two to four pages so that the book can be picked up and put down easily. Each chapter begins with an overview and ends with a summary section. Helpful illustrations are used throughout.

Books currently in the Shopfloor Series include:

5S for Operators
5 Pillars of the Visual Workplace
The Productivity Press Development Team
ISBN 1-56327-123-0 / 133 pages
Order 5SOP / $25.00

Quick Changeover for Operators
The SMED System
The Productivity Press Development Team
ISBN 1-56327-125-7 / 93 pages
Order QCOOP / $25.00

Mistake-Proofing for Operators
The Productivity Press Development Team
ISBN 1-56327-127-3 / 93 pages
Order ZQCOP / $25.00

Just-In-Time for Operators
The Productivity Press Development Team
ISBN 1-56327-133-8 / 84 pages
Order JITOP / $25.00

TPM for Supervisors
The Productivity Press Development Team
ISBN 1-56327-161-3 / 96 pages
Order TPMSUP / $25.00

TPM Team Guide
Kunio Shirose
ISBN 1-56327-079-X / 175 pages
Order TGUIDE / $25.00

Autonomous Maintenance
Japan Institute of Plant Maintenance
ISBN 1-56327-082-X / 138 pages
Order AUTMOP / $25.00

Focused Equipment Improvement for TPM Teams
Japan Institute of Plant Maintenance
ISBN 1-56327-081-1 / 138 pages
Order FEIOP / $25.00

TPM for Every Operator
Japan Institute of Plant Maintenance
ISBN 1-56327-080-3 / 136 pages
Order TPMEO / $25.00

OEE for Operators
Overall Equipment Effectiveness
The Productivity Press Development Team
ISBN 1-56327-221-0 / 96 pages
Order OEEOP / $25.00

Cellular Manufacturing
The Productivity Press Development Team
ISBN 1-56327-213-X / 96 pages
Order CELLP / $25.00

Productivity Press, Dept. BK, P.O. Box 13390, Portland, OR 97213-0390
Telephone: 1-800-394-6868   Fax: 1-800-394-6286